AF577220

Agrarian Reform

Agrarian Reform and Peasant Economy in Southern Peru

David Guillet

University of Missouri Press
Columbia & London, 1979

To my mother,
Juanita Simons and
in memory of my father,
Wilber Francis Guillet

University of Missouri Press, Columbia, Missouri 65211
Library of Congress Catalog Card Number 78–19644
Printed and bound in the United States of America

Library of Congress Cataloging in Publication Data

Guillet, David.
Agrarian Reform and Peasant Economy in Southern Peru.

Bibliography: p. 213
Includes index.
1. Land reform—Peru. 2. Agriculture, Cooperative—Peru. 3. Peasantry—Peru. I. Title. II. Series: University of Missouri—Columbia.
HD556.G8 333.1'0985 78–19644

Acknowledgments

Financial aid for this study came from a number of sources. Travel to Peru and maintenance during 1971–1972 was provided by a grant from the Institute of Latin American Studies of the University of Texas. It was supplemented, later, by a renewal of the same grant and a grant from the Wenner-Gren Foundation to hire research assistants. I returned to Peru for a month in early 1977 as part of research into production organization of the peasant and reform sectors. During 1976 and 1977, I was the holder of a Rockefeller Foundation Postdoctoral Fellowship, attached to the Interamerican Institute of Agricultural Sciences of the Organization of American States. Secretarial assistance along the way was provided by Corinne Bybee and Laura English. The Mary Ashby Cheek fund of Rockford College provided financial assistance in rewriting the original dissertation on which this study is based.

My intellectual debts to faculty members at the University of Texas who supervised my training in anthropology are numerous. Morgan Machlachlan first introduced me to economic anthropology and its relation to development and encouraged me in setting my research within that broad area. Richard N. Adams stimulated in me a deep interest in Latin American social anthropology; his theoretical contributions and insights have permeated my thinking and the pages that follow. Lastly, Richard P. Schaedel, both formally, as the chairman of my dissertation committee, and informally, as friend and mentor, has through his constant intellectual and emotional support, furnished the motivation to see this study through to its completion.

To adequately acknowledge all of the individuals and institutions who assisted me in my stay in Peru would be impossible. Special thanks are due to Ron Skeldon, a geographer from the University of Toronto, who shared lodgings with us at the Jacinto Cáceres's home in Cuzco. Ron had been in the country prior to my arrival and was of great assistance in

Cuzco and Lima in my initial contacts with the various official agencies and individuals attached to them. In Lima, the Centro de Investigaciones y Capacitación de la Reforma Agraria made its library available to me. The Instituto Geográfico Militar was a particularly valuable source of maps and aerial photographs. In Cuzco, my association with the Anthropology Department of Universidad San Antonio Abad as visiting researcher, and, in particular, with Oscar Nuñez del Prado, Jorge Flores Ochoa, and Demetrio Roca Walparimachi, was provident in a number of ways. They were able to recommend two well-trained anthropology graduates, Elisabeth Kuon Arce and Ricardo Cornejo, who proved to be excellent field assistants. A third, Berta Galvez Montero, was trained at the Catholic University in Lima.

I am especially thankful to the many residents of the three communities in the Pampa de Anta who took us into their homes, answered our endless questions, and displayed an amazing understanding of our occasional difficulties. Although they shall remain anonymous to the readers of this study, they are alive in our memories and warm in our hearts. I also wish to thank my friends Dr. Jacinto Cáceres (now deceased) and his wife Amparo Cáceres, who helped with my Quechua and guided me through various legal entanglements.

My wife, Nancy, and I were a team in the truest sense of the word: sharing hardships and rewards equally and providing the mutual emotional support necessary in such a venture. Without such a team effort, the study that I now place before you, would probably never have come to fruition.

D. G.
Kansas City, Missouri
July 1978

Contents

A view from Rumipata of the Pampa de Anta

The *feudatario* community of Tukiwasi and the surrounding hacienda lands

A peasant family from Rumipata

An *ayne* work group

Spraying insecticide on a cooperative potato field

Spinning and weaving

A *tienda* operated by a return migrant from Lima

List of Figures

La reforma agraria requiere la creciente
y libre participación de los campesinos.
Para ellos se hizo, y ellos deben ser
los actores principales del proceso.

Juan Velasco Alvarado

The agrarian reform requires the growing and free
participation of the peasants.
It was made for them and they should be
the principal actors in the process.

1

Introduction

To the observer of Latin American affairs, the demeanor of the military government that came into power by coup in Peru in 1968 is particularly interesting. Peru is no different from most Latin American countries in that the military has been a traditional support of the reigning political elite. Nonetheless, when the military came into power in 1968 it succeeded in enacting, by decree, basic institutional reforms that have reached to the heart of the Peruvian system. Not only were these decrees enacted, but they were acted upon, causing many observers, including Fidel Castro,[1] to endorse Peru as a model that could be profitably emulated by other countries in Latin America.

The number of reform measures carried out by the military is impressive. They include a new code governing the contractual relations between Peru and foreign enterprises, a new law on water control, a new education law stressing bilingual and bicultural education, a reorganization of the State banking, tax, and credit systems, and a reorganization of the government bureaucracies concerned with rural and urban development. Perhaps one of the most unexpected, given the traditional relation between the military and the agrarian elite in Peru, has been the agrarian reform.

The agrarian reform law passed by the military junta in 1969 attempted to resolve the traditional inequity of land tenure in Peru that was felt by many to be a major barrier to the socioeconomic development of both the peasant sector of the population and the nation as a whole. This pattern was most apparent in the sierra region which along with the coast and the tropical *selva* make up the three natural eco-regions into which Peru has been traditionally divided. Particularly in the southern part of the sierra, a small percentage of semi-

1. From the *Economist para América Latina* (London), 23 July 1969, p. 10.

resident landowners, referred to as hacendados, owned a majority of the cultivable land, while a much larger percentage of peasant holders of small- and middle-sized plots of land controlled the remainder. Such a "top heavy" pattern of land tenure was associated with the concentration of power and prestige in the hacendado class. In the national context, the sierra has been a relatively unproductive region lacking an industry that could provide manufactured goods for national consumption and failing in the agricultural sector to provide sufficient produce of high quality for consumption in the coastal cities or for exportation to international markets. There are exceptions, such as parts of the central sierra or the Convención Valley in Cuzco Department, but they are of recent origin and in response to a set of peculiar historical circumstances. In the sierra, the 1969 agrarian reform was designed to redistribute land in a more equitable manner and thus increase the contribution the region makes to the agricultural sector of the national economy.

The planners of the agrarian reform evaluated several other cases of agrarian reform in Latin America in terms of their ability to meet the goals of increased economic productivity and the incorporation of "marginalized" segments of national populations. On the one hand, there were the more conservative cases of Venezuela, Colombia, and the earlier Peruvian attempt at an agrarian reform carried out by President Fernando Belaúnde; on the other hand, there were the examples of Bolivia and Mexico. The differences between the two extremes and the relative advantages and disadvantages of both were available for the military junta to consider. This is apparent in the conscious attempt by the planners to avoid some of the more obvious problems that had weakened earlier reforms, most notably the slowness with which the conservative agrarian reforms were carried out and the subsequent ease with which they got immersed in government bureaucracies. The lack of technical assistance to beneficiaries has also been pointed out as a failure in many of the projects. And, as the more radical reforms have shown, rapid land distribution without minimum planning often results in considerable loss of time and income as the national economy and the individual beneficiaries adapt to the new economic situation. In an underdeveloped country, where agricultural weaknesses often necessitate the use of scarce foreign exchange to purchase agricultural foodstuffs from abroad, a drop in production levels has obvious economic and political correlates.

The military junta has attempted to compensate for the problems that plagued earlier agrarian reforms. First of all, the law represents a serious attempt to resolve directly the most immediate problem of the traditional socioeconomic system by the redistribution of all land above the minimum that ranges from fifteen to fifty-five hectares according to the principle that the land belongs to him who works it. The law has been popular with the middle-class supporters of the military junta who have no stake in preserving the status quo and in fact identify with the peasant, seeing the struggle in the Andes as a symptom of Peru's roots in the past. Second, the law has been well publicized and made a tremendous impact with the expropriation of the highly productive sugar plantations of the north coast within twenty-four hours after the law had been declared; following the coastal actions the agrarian reform moved into the sierra with a well-reported program of planned expropriations. And, lastly, the reform tried to hedge against the possible negative economic effects of the policy by including provisions for the cooperative management of expropriated land, varying according to the socioeconomic characteristics of each region, technical assistance, and a system of priorities of the order in which regions and types of tenure situations would be affected.

Aside from the implications that redistribution of land might have for the economy of the sierra, the agrarian reform has also affected the social system. To understand these changes we may begin initially by dividing the social system into three groups. The first is a large, indigenous population characteristically illiterate and displaying varying combinations of "Indian" traits. A second smaller mestizo population is primarily urban based, which more clearly emulates Peruvian "national" society, and through its ties with the hacendado class controls the sources of power and prestige within the region. The third group acts as an intermediary in the flow of information between the other two groups. Depending on the microregional context, this group may be referred to as *cholo*, mestizo, *misti*, or a number of other terms. As Fernando Fuenzalida has pointed out (1970:66–70) the difficulty of defining this group in class terms stems from the lack of clearly definable indicators that may discriminate members of this group on more than a microregional basis. The intermediary group is best seen as operating in the flow of information between members of rural indigenous communities[2] and the larger society. Often the role of this group in the exchange of information is correlated with the function of facilitating the exchange of material resources between different components of the national society.

A number of changes in the set of social relations, which I have briefly outlined, occurred following the implementation of the agrarian reform. These include new alignments in the relative position of each of the groups and important changes in the makeup and functioning of the intermediate groups. From this viewpoint, the agrarian reform is an economic policy with strong social overtones. This dual aspect of the agrarian reform has not been ignored by the military junta, and as many of their statements indicate they are quite specific in attempting to incorporate the marginalized sectors of the Peruvian population into national life through policies such as the agrarian reform.

This work is a study of the participation of peasant beneficiaries in a cooperative formed as part of the agrarian reform in a microregion, the Pampa de Anta, located in southern Peru. Participation is defined here as joining and becoming involved in the economic, social, and political processes embedded in the activities of the cooperative.

The question of participation is important in relation to the ability of the cooperative to meet the larger goals of the agrarian reform. As an economic unit, supplanting the inefficient hacienda, it strives to allocate efficiently its factors of production, including land, labor, and capital, in maintaining or raising pre-reform levels of productivity. Inasmuch as there has been concern with rural-urban migration and its assumed relation to a surplus of peasant labor in the countryside, the cooperative provides a demand for labor and in extension an

2. I am using *rural indigenous communities* as a broad term to refer to any type of rural peasant community located in the sierra. It is not to be confused with the older term *comunidad indígena* (Indian community), which has a specific legal and political meaning stemming from Peruvian legislation.

alternative to migrating to the city in search of employment.[3] Thus, if successful, the cooperative would increase the contribution the sierra makes to the agrarian sector of the economy.

Participation is social and political as well. The traditional agrarian structure allocated political power to those groups and to their representatives who maintained control over land. Giving some of this control to the peasants and bringing them into the regional and national social and political systems is an important objective of the agrarian reform. Through provisions for the articulation of peasant interests, the cooperative can be an important vehicle toward these ends.

To understand the process of participation of peasant beneficiaries in the Tupac Amaru II cooperative, the anthropological perspective is used. The anthropological perspective implies several things. First of all, it means that use has been made of one of the stronger contributions of anthropology to social-science methodology, the close contextual analysis of individuals and groups in their natural environments. Thus, the data reported here come from fieldwork done in peasant communities in Peru in 1971–1972 using traditional anthropological techniques of participant observation, censusing and surveys, and other techniques described in the Appendix. Through close and intimate contact with the beneficiaries, peasants living in communities and immersed in their day-to-day activities, a version of the impact of the agrarian reform can be presented that often escapes the policymaker.

One of the criticisms, though, of this perspective is that it enables anthropologists to hide behind microstudies that are not representative of the larger whole. Certainly this criticism is valid of the micro-macrocosmic analogies that have been made in the community study school of anthropology and criticized by various writers (Anthony Leeds 1973; Margaret Stacey 1969). Thus to have spent the entire field period in a single peasant community attempting to understand through it a process common to all peasant communities in Peru would have been misleading to say the least. As in other parts of Latin American (David Guillet and Richard P. Schaedel 1973),

3. This is assuming a strict economic motivation behind rural-urban migration. While economic motives are not the only motives behind migration in Peru (Jorge A. Flores Ochoa 1971; David Guillet 1976), they usually are perceived as such by agrarian planners.

there are various types of peasantry in Peru, each of which has different interests and different reactions to a multifaceted policy such as the agrarian reform.

To compensate for the tendency to overgeneralize from a single unrepresentative study of a community, some attempt was made to control the various types of peasantry encountered in the study region. Primary data come from fieldwork in three peasant communities in the Pampa de Anta. These communities were selected to represent (1) communities ranked along a continuum from poor to rich, as measured by cultivable land per household, with the expectation that, all things being equal, participation of peasantry would vary according to wealth as a measure of security, and (2) to represent basic types of communities along a continuum peculiar to the Pampa de Anta of the degree to which peasants display dependence on the hacienda. The poles of the continuum are the hacienda with a resident *feudatario* population and an independent community with no ties to the hacienda. The continuum and the manner of selection of the communities are presented in greater detail in the next chapter and the Appendix.

Anthropological analyses have traditionally been "holistic" in approach. *Holism* means that the unit of analysis, whether it be the tribe, band, or primitive state, is perceived as a whole made up of parts subject to anthropological analysis. In fact, the history of anthropology can be understood by tracing the various holistic models of socio-cultural units. These models specify the types and relationships of components of the socio-cultural unit. Some examples are the well-known models of culture as aggregate of traits, organism, and ecological system.

Holism "worked" best in analyses of isolated tribes and bands and even the primitive state where the boundaries of the unit were usually discrete and observable, and served to isolate a socio-cultural whole with a large degree of autonomy. Such isolation, though, can no longer be assumed for peasantry, which is linked to the nation-state through a number of mechanisms such as the market, taxation, and the networks of communications and transportation.

The problem we are faced with here is multifaceted: to adequately treat peasants as a segment of a nation-state, to treat other segments, such as the nation and the region, that are instrumental in the relationships with peasantry, and, lastly, to deal with the processes through which all segments interact as component parts of a complex society. One direction taken in this study is to discriminate levels of societal integration. The concept of level has been found useful as a way of ordering the diversity of socio-cultural units that make up complex societies. Julian Steward first used the concept in 1951 and later developed it in an evolutionary theory of cultural change (Steward 1955). He dealt with principally two levels in complex societies: the community, or local level, and the national level (Steward 1955:55). Eric Wolf has carried the concept further in referring to a series of levels in Mexico in his analysis of the breakdown of Indian communities and haciendas. His rationale for using the level is that it "illuminates the manifold processes of conflict and accommodation which take place when the components of a socio-cultural system are rearranged to answer new needs or taken up into more embracing systems" (Wolf 1967:299–300). Wolf discovered a series of levels including the nuclear family, kindred, barrio or ward, community, constellation of town center with dependent communities, constellation of regional capital with satellite towns, and finally the State.

Three levels have been chosen, the national, the regional, and the community or local level, as basic organization and analytical units. These levels represent constellations of actors that have been instrumental in the agrarian reform process. The agrarian reform emerged out of a series of events in the national social structure of the country that culminated in the military coup of 1968 and the subsequent reformist orientation of the government. It is essential to understand this process in order to analyze the form that the agrarian reform eventually took. These events are analyzed in Chapter 4; in particular, the chapter will look at the cooperative concept as a leitmotif that permeates all aspects of post-1969 governmental reform policy. To administer the program, the country was broken up into agrarian reform zones; within zones, regions were defined and assigned priorities with respect to the order of implementation. Understandably, these administrative units corresponded to natural geographical regions and somewhat less closely to political divisions such as Department, Province, and District. Our usage of region will refer to the Department of Cuzco and the capital city of the same name; here we find the offices and the administrative center of Agrarian Reform

Zone Twelve. The regional level also includes the Pampa de Anta, a natural microregion, which is the site of the three communities studied during 1971 to 1972. In Chapter 2 an ethnographic overview of the Pampa de Anta is presented: its locational, ecological, demographic, political, and social aspects. The three communities upon which much of this study is based are placed within these multiple contexts.

I have chosen to regard the Pampa de Anta with Cuzco as a unit because there is an important continuity in relations between the mestizo class of the Pampa and the agrarian elite of Cuzco. Mestizos in the Pampa have traditionally wielded power allocated to them by an agrarian elite resident in Cuzco. The political economy of the Pampa reflects the coalescence of their interests. Thus, in Chapter 5 we will discuss the series of moves associated with the implementation of the agrarian reform in the Department of Cuzco and the Pampa de Anta. These include the growth and development of an agrobureaucracy that replaced the agrarian elite and its dependent mestizo class, the expropriation of haciendas, the distribution of land to the Tupac Amaru II production cooperative, and measures taken to promote the cooperative to the peasant beneficiaries. In Chapters 6 and 7 we will move to the local level and explore the implementation of agrarian reform policy in the three communities. Chapter 6 will focus on the economic organization of the cooperative as contrasted with the communities, while Chapter 7 does the same for the sociopolitical aspects of the cooperative. Some conclusions derived from the study are presented in Chapter 8. Developments in the period from 1972 to 1977 are discussed in a postscript based on a visit to Peru in 1977.

To understand the process of participation we have drawn on a trend in contemporary socio-cultural anthropology that looks at the situational context within which behavior occurs. In such an approach, the interest is in why people make the decisions they do in a given context. It focuses on the constraints people are faced with and the strategies they utilize in making decisions. Participation, indeed, involves a number of strategic choices: beneficiaries choose to join the cooperative; collective decisions regarding agrarian policy are made by peasants in community assemblies; local assemblies of cooperative members arrive at decisions concerning issues of cooperative policy; members decide whether to work on cooperative land or their own, and so on. An inquiry into these and other decisions will reveal constraints to full participation; combined with a multilevel analysis, we can further locate these constraints at the level of society in which they originate.

We can draw a contrast between the anthropologist's attempt to deal with behavior at the individual level and other modes of analysis in the social sciences. The contrast is more apparent with, what might be called, macro-models of human behavior constructed by economists, political scientists, and sociologists. Macro-models focus on aggregate data and give a statistical description of the phenomenon under study. In this case there has been a tendency to "black box" that part of the system to which the methodologies of the investigator are unable to speak. Microanalyses of individual behavior as done by anthropologists complement macro-models in revealing the full play of variables that are incompletely understood, overemphasized, or at worst ignored, by macro-models.

An example of the kind of process associated with participation that can be illuminated through micro-level analysis is that of joining the cooperative. During the first year of operation of the cooperative, less than half of the qualified peasants joined (Carlos Kawata et al 1972:59). Fifty-one percent of these were *feudatarios,* peasants who rarely owned land of their own but who exchanged labor for a plot of land owned by a hacienda. *Feudatarios* usually live on the land of a hacienda and less often in adjoining independent peasant communities. When these haciendas were expropriated by the agrarian reform, the *feudatario*'s land became part of the cooperative. They had little choice but to join the new cooperative in order to maintain access to land. There remained a large pool of peasants living in independent communities, though, that were offered the option of joining the cooperative, but who responded to a much lesser degree than the *feudatario* group. Thus, aggregated data on the composition of the membership of the cooperative would reveal the relative contribution that each group makes to the total membership. But the process, which includes perception of the cooperative as an option, can best be understood through close contextual analysis of individual and group behavior.

One of the variables that will be given particular attention is the amount of information that is available about the option to join and participate in the cooperative. The amount of information is extremely important in the type of problem that the decisionmaker confronts. Decisions can be ranked according to the amount of information that is available:

> Certainty if each action is known to lead invariably to a specific outcome. . .risk if each action leads to one of a set of possible outcomes, each outcome occurring with a known probability. The probabilities are assumed to be known to the decision maker. . . uncertainty if either action or both has as its consequences a set of possible specific outcomes, but where the probabilities of these outcomes are completely unknown, or are not even meaningful. (R. Duncan Luce and Howard Raiffa 1957:13)

Ordinarily, peasants can be said to follow strategies in which production factors are allocated in combinations that are low in risk and low in return.[4] This is a reflection of the nature of the peasant economy. Often having to eke a subsistence from land that is marginal in quality and insufficient in quantity to meet the needs of an ever-growing household, the peasant is particularly sensitive to a number of factors over which he has no control; climatic factors, for example, such as rain, drought, hail, and frost; as well as societal factors such as the insecurity of living in a national context where socially, economically, and politically, the peasant must obey the dictates of members of the larger society. Under these conditions peasants are extremely concerned with security; a low risk-low return strategy is a reflection of that concern.

The production cooperative that was organized in our study region as the second stage of the agrarian reform, following the expropriation of the majority of the haciendas located in the region, in effect, offered an economic alternative to the rather limited set of "income streams" that had been exploited by peasants in their traditional agricultural setting. The concept "income stream" implies, following Theodore

4. Another way of characterizing the situation is to say that peasants are concerned with maximizing security, that is, the provision of food, water, shelter, and clothing. Johnson, in describing Brazilian peasants' economic behavior, says: "In their agriculture, and in their social relations, their behavior is directed almost entirely toward meeting basic subsistence needs; they do not do things in a certain way 'because things have always been done that way,' but rather because this way will work reliably to assure survival" (Johnson 1971:132).

W. Schultz (1964), an income-producing commodity with a cost and an empirically definable return. Joining the cooperative gives the member access to a number of different types of income streams such as the right to pasture one's cattle on cooperative pasture lands and a relatively high salary paid in return for working for the cooperative. The salary of twenty-four soles per day paid by the cooperative is a considerably higher sum than the rate available to most peasants working on a day-labor (*jornal*) basis, varying from ten to fifteen soles per day.[5]

The problem, as interpreted in this study, lies in the perception of the income stream by the peasant beneficiaries. Since the cooperative is new, and knowledge about its functioning is scarce, the prospective member is considerably uncertain about the outcome of his decision to join the cooperative. If he were certain or at least reasonably certain, then his decision would be much easier to make, and from what we know about the cost and return to be had, we would comfortably predict that a prospective member would join the cooperative and thus "buy" the income stream.

The Peruvian planners were quite aware of the necessity to disseminate information concerning policy and designed a communications program as part of the implementation of the agrarian reform. This program was based to a large extent on the model of the diffusion of innovations propounded by Everett M. Rogers and his followers (Everett M. Rogers and Lynne Svenning 1969). An expansion of the agrobureaucracy, the use of group discussions, a radio forum program, and change agents were some of the techniques employed by the program. The objective of the program was to provide a set of information sources available to the peasant under government auspices.

Thus, in this study I will evaluate the extent to which the government program provided an adequate amount of infor-

5. This represents the rate that prevailed during the period before the agrarian reform. Subsequent legislation set twenty-four soles a day as a minimum salary and began to be the minimum that was paid by the middle-sized landholders (*pequeños propietarios*), especially since they were motivated by the rumors that failure to pay the minimum wage would result in expropriation of middle-sized landholdings. Ten to fifteen soles, though, were paid to peasants recruited by the small-holding classes of peasants.

mation about the inner workings of the cooperative, through the perspective of the peasant beneficiary. There is another, perhaps more subtle, side to the question of the availability of information, though. One of the principal findings of anthropological research is that information is highly culture specific and must be relevant to the context of the individual. This would include both the form that the diffusion of information takes as well as its content. A moment's reflection on the cooperative should illuminate the problem. The option to join the cooperative implies acceptance of a model of production containing a number of elements that can be assumed to differ substantially from a non-Western indigenous culture such as the Quechua. Concepts such as "shares," "delegates," "patrimony," and "commercialization" are used to describe the resources manipulated in the cooperative production system. Procedures upon which production decisions are made are quite possibly unique to the cooperative system; the peasant is told, for example, that he is to "participate" in the decisions of the cooperative through a series of "levels" of participation. On the local level members elect "delegates" that represent them at the higher levels. At the highest level—the "general" assembly that includes all of the delegates—the "annual work plan" is formulated and agreed upon. The annual work plan is then given to an individual who is hired to carry it out on a day-to-day basis. Aside from the strategies associated with the allocation of resources in the cooperative, the resulting lack of correspondence on the conceptual level between the native and the introduced system is a good example, although often ignored by development economists, of a type of "cognitive dissonance" between two conflicting cultural systems.[6]

6. A good example of lack of correspondence on the conceptual level is Acheson's (1972) discussion of Tarascan accounting concepts and those derived from formal economics. He suggests that the formal economic concepts of accounting confuse the perception of opportunities by Tarascans and lead to many poor business decisions.

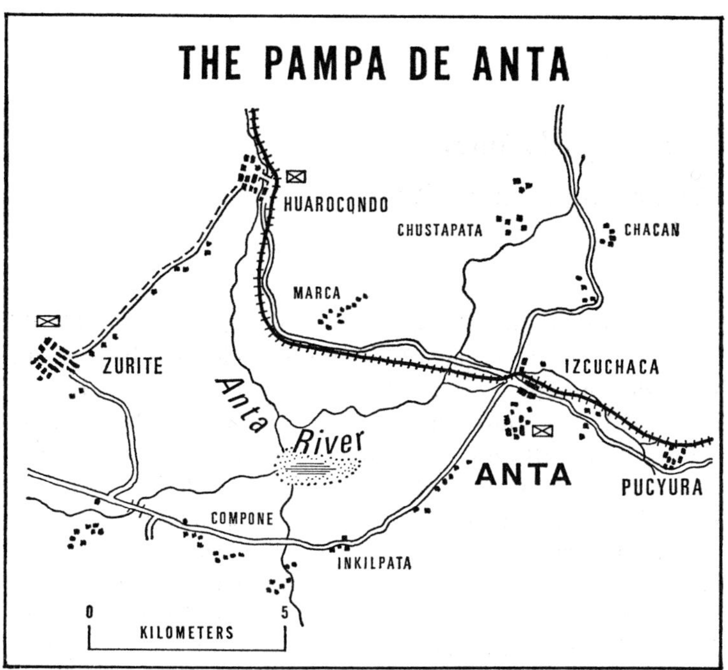
THE PAMPA DE ANTA
HUAROCQNDO
CHUSTAPATA
CHACAN
MARCA
ZURITE
IZCUCHACA
Anta
River
ANTA
PUCYURA
COMPONE
INKILPATA
0
5
KILOMETERS

2

The Pampa de Anta: Region and Community

The three communities, Antapampa, Rumipata, and Tukiwasi,[1] that are the object of this study are located in a natural ecological microregion, the Pampa de Anta, in Cuzco Department of Peru's southern sierra. The Pampa de Anta, or the Pampa as it will be referred to, is one of the many highland basins formed by rivers draining branches of the Cordillera, or major branch of the Andes that flows northeast through Peru. The Pampa is a product of the meeting of the headwaters of the Apurimac and Urubamba rivers bordered by mountains reaching around 21,000 feet in altitude. Its floor is a high flat plain 10,000-10,200 feet in altitude, 7½ kilometers in length, and 5 kilometers in width at its widest point.

The Pampa is intersected by a major highway connecting Cuzco and points south with Lima; another highway crosses the Pampa and unites Cuzco with the Urubamba and Convención Valley to the east. These roads provide constant opportunities for contact between the Pampa and other major population centers in the northern and southern sierra, the tropical forest region to the east, and ultimately the coast. In addition, the Cuzco-Quillabamba railway crosses the Pampa with stops at Izcuchaca and Huarocondo. The major communications point in the Pampa is Izcuchaca adjacent to Anta where the northeast and north-south roads cross. Trucks stop there on their way to Lima and Quillabamba; buses connecting Cuzco with the Pampa stop there. Also, a market operates throughout the week, and considerable commerce takes place. Anta is the capital of the province of the same name.

Zurite, Anta, and Huarocondo are *aldeas* or capitals of the districts of the same name. In relation to size and function, they are the lowest rung of the Peruvian urban hierarchy. Their economy has depended until recently upon the market-

1. These names are pseudonyms.

ing of produce of the countryside. In each *aldea,* a periodic market brought peasants to sell or exchange their produce with resident or Cuzco based middlemen. In return they purchased produce from other regions and a limited number of manufactured items such as rum, sugar, matches, candles, carbonated beverages, and the like. The livelihood of the *aldeas* depended almost entirely on this commerce rather than any explicit function arising from their position in the political-administrative network. In practice, most political administration in the Pampa is centralized at the level of the Provincial capital, Anta/Izcuchaca. Recently, the economic base of the *aldeas* has been considerably weakened as peasant markets have sprung up at different points in the Pampa. Rumipata, for example, had a market to serve the needs of the residents of the high puna communities. When a road to Cotabambas was constructed in 1963–1964, the market ceased to function. Ancawasi is now the locus of a market that functions in many ways like that of Rumipata, as a channel for the exchange of goods from two ecological zones. In both cases, peasant middlemen now perform an entrepreneurial function more successfully than the "urban" merchants of the *aldeas.*

Outside the *aldeas* and the major communications center at Izcuchaca, the real life of the campo begins. The majority of the residents of the countryside are peasants, descendents of an indigenous Quechua-speaking population who were the inhabitants of the region at the time of the Spanish Conquest. The community is the pervasive unit of the countryside easily distinguished in some cases by a compact settlement of thatched roof, adobe houses, and a church or school nucleus (*villorio* or *villorio nucleado*), and more difficult to distinguish in other cases where houses are dispersed with no apparent order (*caserio*). Each community contains a grouping of peasant households that maintain a common identity based on residence in a locality defined in territorial terms. Further, each community has a peculiar mix of organizational features: kinship is relatively weak as an overall mode or organization, but the more fluid types of kin and quasi-kin bonds such as *compadrazgo* or fictive kin, friendship, and the kindred based on residential proximity, are quite strong in organizing peasants within a community. Another type of organizational format is along political lines; some type of decisionmaking body, based either on power allocated to a set of officials drawn from the members of a community, or power delegated from extra-communal power sources to representatives in a community, is found. The decisionmaking body functions internally, as an equilibrating mechanism and conflicit-resolving device, and externally in the articulation of community needs and defense.

Horizontal class divisions exist to varying degrees in all of the communities; these divisions are a reflection of a number of processes, including differential access to economic resources, acculturation, and a distinction between groups oriented to cultural influences stemming from the "national" culture and groups essentially inward and community oriented. The latter distinction implies that the described peasant communities are part of a larger, and in many ways different, cultural matrix; this matrix is at present the Peruvian "national" culture in its many forms. From the perspective of the majority of the peasants in the Pampa, the matrix maintains a constantly changing form and content, radically different from lifestyles or patterns to be found in the community. The mediation between nation and community is a major source of tension and conflict in the community that manifests itself in many different ways.

The picture outlined here corresponds to other peasant societies throughout the world (Janet M. Fitchen 1961). From a structural viewpoint, in terms of the economic links with the larger society, there are some important distinctions among the range of communities in the Pampa. These distinctions refer to what may be termed the *regimes* of these communities: basically, we may distinguish two regimes, with a series of variations. Most simply, there is a distinction between the hacienda community and the independent community. The hacienda community is a residential grouping of peasants most of whom have entered into a contractual relationship with a hacendado, or landowner, in which they exchange labor for a subsistence plot of land and a minimal wage. Peasants entering into this relationship are called *feudatarios*. The hacienda is a result of the alienation of land from Indian communities since the Conquest, and its subsequent concentration in the hands of a owner through manipulation of the juridic-political institutions. In the independent community, on the other hand, peasants own land either independently or as a result of their being members of the community. In the latter case, the community is a kind of corporate landholding unit that allocates and redistributes land to its members under a number of forms. According to a census made in 1970 by the Peasant Communities Agency (Ministerio de Agricultura, Dirección de Comunidades Campesinas 1970:4–7), there were thirty-seven independent communities and seventy haciendas in the Pampa de Anta. Twenty-three of the independent communities were formally recognized under provisions of the 1920 Peruvian constitution. While giving some legal protection against encroachment on communal lands, this status does not imply any major socio-cultural difference between recognized and nonrecognized independent communities. Thirty haciendas of the total were rather small operations whose labor force was made up of peasants living in adjoining independent communities and who constituted a special category. These variations in political status, land use and control, and mode of relationship to the hacienda will be discussed in greater detail.

While Rumipata and Antapampa have large collections of houses dispersed over the landscape with no apparent order nor central focus, Tukiwasi, on the other hand, is a highly nucleated cluster of houses located around a central plaza with a church at one end. There are no other features to distinguish them, and to the passerby they look the same as all of the other similar communities on the Cuzco-Lima highway. Even the most important characteristic of these communities—that one (Tukiwasi) is a hacienda, and the others are autonomous peasant communities—is not readily apparent to the observer.

Population, Land, and Growth

As in other parts of Peru, one of the major factors affecting the structure of *aldeas*, haciendas, and communities is their rapidly changing demographic composition. Figure 2–1 presents population figures for each of these units in the Pampa de Anta taken from the 1940 and 1961 censuses. *Aldeas* decreased considerably in reference to population during this period with Zurite and Huarocondo sustaining the greatest loss. This is a reflection of the point made earlier that the economy of *aldeas* has depended on a continual flow of produce from the countryside to merchants and middlemen

Figure 2–1. Population Dynamics in the Pampa de Anta, 1940–1961

	1940	1961	Net Growth 1940–1961	Rate
Aldeas	8,892	7,006	–1,886	–21.2
Anta	1,542	2,574	1,032	66.9
Huarocondo	5,572	2,921	–2,651	–47.6
Zurite	1,778	1,511	– 267	–15.0
Haciendas	1,836	2,077	241	13.1
Communities	8,618	12,175	3,357	41.3
Pampa de Anta	28,418	26,041	–2,377	– 8.4

Source: Peru, Dirección Nacional de Estadística, 1944; Peru, Dirección Nacional de Estadística y Censos, 1966.

located in the towns. The recent shift from markets located in Zurite and Huarocondo to more optimally located markets in the countryside based on peasant entrepreneurial middlemen to a large extent has brought about their plight. They are essentially stagnating, unable to attract or keep a population because of the lack of a dynamic economic base. Anta's unusual growth can be attributed to its proximity to Izcuchaca, the communications and transportation center of the Pampa. Haciendas and communities fared much better in retaining their population and grew considerably during 1940 and 1961.

Turning to the three communities we find the rather unusual growth pattern presented in Figure 2–2. Rumipata has grown steadily since 1940 achieving the high 44.9 percent growth rate for the latest census period. Tukiwasi grew considerably during the first period but began to slow following 1961 and actually lost population by 1970. Antapampa has the most spectacular pattern of the three communities.

Figure 2–2. Population Dynamics in Rumipata, Antapampa, and Tukiwasi, 1940–1970

	1940	1961	Net Growth 1940–1961	Rate	1970	Net Growth 1961–1970	Rate
Rumipata	510	666	156	30.6	965	299	44.9
Antapampa	103	695	592	574.8	667	–28	–4.0
Tukiwasi	134	241	107	79.8	250	– 9	–3.7

Source: Peru, Dirección Nacional de Estadística, 1944, Peru, Dirección Nacional de Estadística y Censos, 1966; Peru, Ministerio de Trabajo, 1970.

Following close to a sixfold increase in size during the first census period, 1940 through 1961, the community slowed dramatically and actually lost population by 1970 the next census period. The spectacular growth of Antapampa during the 1940s and 1950s seems to be tied in with the growth of Izcuchaca; Antapampa is only a ten-minute walk from Izcuchaca and is the first major stop on the highway leaving Izcuchaca for points north. The growth pattern is reflected in numerous homes that have been built next to the highway and away from the fields where homes have been traditionally dispersed. Informants state that many families moved to the community in order to be close to the services in Izcuchaca: commercial activities, a school serving the secondary grades, and the trucks and bus lines that stop there. The decline in population can only be explained as out-migration, which will be discussed later in this chapter.

Another insight into more recent growth patterns is presented in Figure 2–3, which gives household growth patterns

Figure 2–3. Household Growth in Rumipata, Antapampa, and Tukiwasi, 1961–1970

	Year	Enumerated Population	No. of Households	Percentage Increase 1961–1970	Persons per Household
Rumipata	1961	666	132	65.1	5.0
	1970	965	218		4.4
Antapampa	1961	695	160	−10.6	4.3
	1970	667	143		4.7
Tukiwasi	1961	241	52		4.6
	1970	250	53	1.8	4.7

Source: Peru, Dirección Nacional de Estadística y Censos, 1966; Peru, Ministerio de Trabajo, 1970.

for the three communities derived from the 1961 census and the 1970 Peasant Communities Census. While Tukiwasi remained stable in all indicators during this period, household growth accentuated the trends noted in the previous table. While the total number of households declined considerably in Antapampa during this period, in Rumipata, new households were started at a faster rate (65.1) than the population increased (44.9) during the same period. Again the explanation for Rumipata's growth must be sought in external factors: the location of a school in the community, a road constructed to Cotabambas during the 1960s, and proximity to the highway, and other factors, which resulted in a stream of in-migrants contributing to the growth manifested in the tables.

The relationship between the population and land factors must be considered if we are to understand the problem discussed in this work. The general growth trend for haciendas and communities indicates increasing population pressure on land. The land available to peasants living in both haciendas and communities has remained relatively stable; if anything the tendency has been for it to decrease in this century due to encroachment by haciendas. Land distribution in Rumipata, Antapampa, and Tukiwasi will be analyzed in depth in the next chapter but the dimensions of the problem are given in Figure 2-4, which is derived from the Peasant Communities

Figure 2–4. Distribution of Irrigable and Seasonal Land, by Household (In Hectares)

	Irrigable Land	Hectares Household	Seasonal Land	Hectares Household
Rumipata	101	0.46	469	2.15
Antapampa	53	0.37	13	.08
Tukiwasi	19	0.36	60	1.09
Universe	3,245	0.63	5,640	1.12

Source: Peru, Ministerio de Trabajo, 1970

Census. Irrigable land refers to the fertile bottomland that is provided with year-around access to water; it is more productive than seasonal land that lies along the sides of mountains bordering the communities. Seasonal land can only be cultivated during part of the year and is quite susceptible to climatic vagaries.

All three communities possess less irrigable land per household than the total census sample. In terms of seasonal land Rumipata and Antapampa are opposite: Rumipata has approximately twice the average while Antapampa has virtually none. Tukiwasi possesses land in both categories; it must be remembered though that Tukiwasi is a hacienda and land is owned by the hacendado and extended in usufruct to peasants in return for obligations of labor on the lands of the hacienda. We may conclude that Rumipata is a relatively rich community in relation to Antapampa and certainly in comparison with Tukiwasi. Furthermore, the location of Rumipata in a protected ravine, *quebrada,* increases the productivity of its seasonal lands, while the seasonal lands of Tukiwasi and Antapampa are much more exposed.

A further insight into the relative wealth of each of the communities is in the economic category of their residents. While the majority of residents of independent communities are engaged in farming plots of land located in their communities, often land is in insufficient quantity to meet the needs of all the residents. This results in the necessity of securing land on adjoining haciendas. These individuals retain rights to communal lands, as well as land in the hacienda, and are referred to as *comuneros/feudatarios.* A fourth economic category (after *comunero, feudatario, and comunero/feudatario*) is the middle holder, or *pequeño propietario,* who does not belong to or does not participate in the political affairs of either haciendas or independent communities. They usually reside on and own land within the boundaries of an independent community or hacienda but often occupy land between the two units. The economic categories of household heads of the three communities are presented in Figure 2–5. A significant number of *feudatarios* and *comunero/feudatarios* in Antapampa is a reflection of the lack of a sufficient amount of land.

Figure 2–5. Economic Category of Household Head in Rumipata, Antapampa, and Tukiwasi

	Comunero	Com./feud.	Feudatario	Pequeño Propietario	Total
Rumipata	216	1	1	—	218
Antapampa	110	21	4	8	143
Tukiwasi	1	18	33	1	53

Source: Peru, Ministerio de Trabajo, 1970

The most important causes of growth are the changes in fertility and mortality trends that have occurred in Peru in this century. These trends are presented in Figure 2–6, which is based on a United Nations study. While birth rates have remained relatively stable, dramatic declines in death rates have resulted in a consistent rise in the rate of natural population increase. The decline in death rates can be attributed to improved sanitation, medical facilities, and living standards.

Figure 2–6. Decline in Death Rates and Stability of Birth Rates in Peru, 1940–1961

Period	Births (per 1000)	Deaths (per 1000)	Natural Increase (per 1000)
1940–1945 (UN estimate)	43.0	24.5	18.5
1945–1950 (UN estimate)	43.0	23.3	19.7
1950–1955 (UN estimate)	43.0	23.0	20.0
1955–1960 (UN estimate)	43.0	19.0	24.0
1960–1965 (UN estimate)	43.0	17.4	25.6
1961 (official census)	45.0	15.0	30.0

Source: See Table 6, p. 82, in Rolland G. Paulston, 1972.

Figure 2–7, derived from the 1961 census and statistics, presents crude birth and death rates for the Anta district. Rumipata, Antapampa, and Tukiwasi are located in this district and record their births and deaths in its offices. The lower rate of natural increase than that for the nation is a reflection of the higher death rates in the southern sierra. Improved medical facilities and sanitation practices are just beginning to disseminate into the region and death rates have not yet lowered as much as the more developed coastal regions of the country.

Figure 2–7. Crude Birth Rate, Death Rate, and Rate of Natural Increase in the Anta District, 1961

Enumerated Population (a)	No. of Births (b)	No. of Deaths (c)	CRUDE RATE Birth (b/a)	CRUDE RATE Death (c/a)	Rate of Natural Increase (b-c/a)
11,790	510	233	43.2	19.7	23.4

Source: Peru, Dirección Nacional de Estadística y Censos, 1966; Peru, Dirección Nacional de Estadística y Censos, 1969.

The differences between growth as predicted by the calculated rate of natural increase and the actual growth as calculated through census data in Figure 2–2 is a reflection in part of the effect of migration patterns. Where census figures reveal a population loss or relative stability as in the case of Antapampa and Tukiwasi then we must conclude that considerable out-migration must have occurred since there is no community specific record of epidemic disease or other demographic calamity to account for the loss. Rumipata on the other hand had a greater growth than that predicted by the rate of natural increase as of 1961. As previously mentioned, there have been a peculiar set of historical circumstances to account for its rather remarkable growth.

Most instances of population loss can be accounted for by permanent out-migration usually to an urban center.

Permanent migration to an urban center corresponds to the decision of an individual to move from his community of origin and to set up residence in an urban center. It is an important decision as many risks make living in the city extremely hazardous for the newly arrived peasant. Informants are quite aware of the risks and continually make reference to the high costs of basic commodities such as food, rent, transportation, and clothing in comparison with the low wages obtained for typical jobs held by *serranos* upon arriving in the urban context. It is important to realize that there is a clear preference for Lima over any other urban center in the country; all of the major decisions to migrate usually are based on Lima as a final destination. This has not always been the case. Informants state that until recently (last ten to fifteen years) individuals left their community to go to Cuzco to live. The Club of Rumipata migrants living in Cuzco stopped functioning about ten to fifteen years ago and peasants no longer regard Cuzco as a particularly appealing place to reside.

In Rumipata, where migration to Lima is quite frequent, there is a special fiesta, the *kacharpari,* which is given prior to an individual's departure. A close relative or friend will accept the role as sponsor and provide the house and refreshments for an all-night private fiesta. The next morning the fiesta party will accompany the person to the road to wait for the truck to take him to Lima. The object of the *kacharpari* fiesta is to send off the person with the well wishes and hopes that he is successful in meeting the risks of living in Lima. It is not given for any other place of destination.

An insight into the destination of out-migrants may be had in Figure 2–8. This table was constructed from the responses

Figure 2–8. Location of Offspring of Peasants in Migration History Sample

	Total No. of Offspring	Cuzco	%	Lima	%	Valleys	%	Other	%
Rumipata	177	16	9.0	59	33.3	2	1.1	1	0.5
Antapampa	102	7	6.9	23	22.5	3	2.9	3	2.9
Tukiwasi	54	3	5.5	11	20.4	1	1.9	2	3.7
Total	333	26	7.8	93	27.9	6	1.8	6	1.8

Source: Migration History Sample.

of peasants questioned in the migration study about the location of each of their offspring. Answers were received from a total of seventy-six household heads.[2] The category "valleys" in the table refers to the Convención Valley, Tambopata, and other tropical valleys to the east that attract a considerable number of seasonal migrants during the off-season of the highland agricultural cycle. As is shown in the table, Lima contains the highest number of migrants with Cuzco in second place.

In a recent series of publications, Ron Skeldon (1976, 1977*a*, and 1977*b*) has analyzed the structure of migration in the Department of Cuzco. According to Skeldon, there are five distinct phases of spatial and temporal mobility that vary as the process of urbanization evolves. In the first phase, where urbanization is incipient, one finds seasonal migration between district and provincial capitals and Cuzco together with permanent movement from Cuzco to Lima. Ultimately, there is a final stage, characterized by permanent migration from the smaller rural communities and district, provincial, and departmental capitals to Lima. According to Skeldon's scheme, Anta communities around fifteen to twenty years ago moved out of stage C or a pattern of local semipermanent migration to the departmental capital to phase D or possibly E of semipermanent and/or permanent migration to Lima.

2. The breakdown by community is Rumipata, 38; Antapampa, 25; and Tukiwasi, 13; with a total of 76.

Sociopolitical Organization

Social organization in the three communities can be described with reference to three spheres. The first is the household that ideally and most commonly is the nuclear, bilateral, family. The household is at once the basic unit of production and consumption, legally autonomous within the political organization of the community, and the building block of the larger, more encompassing, spheres of social organization.

Next is the personal network of social relationships created and maintained by an individual. Because of the nature of Andean society, one finds the personal network, with its characteristic flexibility, in place of the more inflexible forms of social organization such as, for example, the clan or the large corporate descent group. The basic reason for this pattern stems from the environment in which the Andean peasant lives. This environment in its widest sense includes the ecological, economic, political, and social forces that are constantly in flux and that require dynamic responses from peasants. This means that one must keep at hand a personal network of individuals from which one can activate an instrumental relationship. For example, in the economic setting this may involve: recruiting a partner who will agree to share agricultural chores during a season; recruiting a still larger work group to quickly finish soil preparation prior to the beginning of the rains; entering into a sharecropping arrangement with an individual who has surplus land in a strategic location; or making an arrangement through which an individual living in a high altitude settlement will agree to care for one's animals during the dry season. These examples could be supplemented by others drawn from other spheres such as the need to find housing and care for one's children in the city to which they move to attend school.

In the creation of a personal network, one has recourse to a variety of preexisting ties as well as mechanisms for establishing new ones. Preexisting ties include one's kindred and neighbors. The kindred consists of one's consanguineal relatives and those affinal relatives created by marriage. Usually it will overlap considerably with residence, that is, a neighbor will be a member of one's kindred because of a tendency to patrilocal/virilocal residence. The most important mechanism for establishing a lifelong relationship is through *compadrazgo*, or ritual co-parenthood.

The nature of the first two spheres of social organization found in peasant communities in the Pampa de Anta is thus basically the same as that found throughout the Andes. It consists of the bilateral, nuclear family as the basic economic, ecological, and political adaptation and a wider group of kindred, neighbors, friends, coresidents of a community, and *compadres* from which a personal network is selected that can be activated for instrumental ends (Bernd Lambert 1977; Ralph Bolton and Enrique Mayer 1977).

The third sphere of social organization refers to long-standing residential segments of a community that interact on a more recurrent and intimate basis than with members of other divisions or with members of the community at large. There are two such divisions in Antapampa and four in Rumipata. In Antapampa the community is divided into two halves; in Rumipata there are four named divisions with clearly demarcated territorial boundaries. Each of Rumipata's four divisions is correlated with the concentration of surnames (*apellidos paternos*) indicating a patri-bias in post-marital residence. Within recent years these traditional divisions have formed the basis for an internal reorganization for labor recruitment in Rumipata and a dispute over the location of a church in Antapampa. These changes will be analyzed in some detail in the next chapter. As a result Antapampa is now divided into a Antapampa Chico, "little" Antapampa, and Antapampa Grande, "big" Antapampa, while Rumipata is organized into four "sectors" that correspond with the older traditional divisions.

The division of the community into multiples of two and four occurs as a central element of Incan social organization. Inasmuch as these principles continue to order social organization in other Andean rural communities (Salvador Palomino Flores 1971), there is no reason to doubt their importance in the social organization of Pampa communities.

Contemporary political organization in Antapampa, Tukiwasi, and Rumipata is a result of the adaptation of traditional political institutions to extra-communal influences impinging on the community. The interrelated fiesta complex and cargo system are the traditional institutional forms. The fiesta complex emerged during the colonial period as a result of Catholic missionary activity (George Kubler 1946:349). It involved a complex of ritual activity that extended into everyday life through the associations, *cofradias,* which organized and sponsored ritual. The contemporary fiesta complex involves the periodic observances of specific occasions of the Catholic ritual calendar and the patron saint of each community. The observance usually includes a religious ritual such as a Catholic mass and a secular counterpart in the form of dancing, eating, and much drinking. While these are "public" fiestas that recruit participants from the entire community, there are also "private" fiestas marking important life crises such as birth, the first hair cutting, the first Catholic communion, confirmation, marriage, and death. Private and public fiestas alike involve the sponsorship of different aspects of the ritual accompaniment; in the case of private fiestas, individuals ritually sponsor other individuals, thus creating formal ties (of *compadrazgo*) among a number of people that last during the lifetime of all the individuals involved. Public fiestas involve a qualitatively different type of ritual sponsorship referred to as the *cargo* system.

The *cargo* system is an indigenous adaptation of the *cofradía* associations that emerged during the colonial period. In it religious offices, or *cargos,* denoting sponsorship of a particular aspect of a fiesta are occupied on a rotating basis by the men of the community. They are successive in that there is an order in which each *cargo* must be held; generally, each successive *cargo* requires continually greater sums of expenditures. And, as a rule, the more money spent on a *cargo,* the more prestige accrues to the holder. Once an individual has passed all of the *cargos,* he is usually acknowledged with a special status, equivalent to an "elder," in the community.

In the past, the *cargo* system was directly tied to the political organization of the community, such that religious *cargos* alternated with political offices (Fuenzalida 1969:42). There was a close correlation between age and the number of *cargos* held. In practice, this implied that the more prestigious *cargos* were held by the older members of the community. The highest *cargo* holder in the civil-religious hierarchy was in effect the political leader in the community; his authority was symbolized by the silver-encrusted staff, the *vara,* which he carried.

For a number of reasons, the *cargo* system as described is no longer used in the three communities. As in other parts of Peru (Oscar Nuñez del Prado 1973:88–89; Billie Jean Isbell 1971) and in Mesoamerica (Frank Cancian 1965) socioeconomic change has adversely affected its input into community life. There is no pressure to attend the public fiestas and to participate in the *cargo* system that existed in the communities up until approximately twenty to thirty years ago. Now, informants say, an individual who doesn't want to hold a *cargo* can "buy a set of clothes for the Virgin" instead. More significantly, the *cargo* systems that remain are considerably truncated, often by a decision of the community, from earlier versions. In Rumipata, for example, approximately four or five years ago, it was agreed that the two most prestigious and expensive *cargos*, the *alferez* and the *maestricampo*, would be eliminated because of their cost and the fact that no one was willing to hold them.

The lack of interest in the *cargo* system and the fiesta complex can be attributed partly to social change that has affected the communities. Education, military service, seasonal migration, and knowledge of the national culture through the media, and trips to the city have placed the *cargo* system and the fiesta complex in a perspective associated with Indian culture rather than the national culture to which many peasants aspire. Holding *cargos* and all of the associations of the *cargo* system, such as the *bailarines* and their folk dance, does not bring the prestige for the younger peasants that more "modern" indicators, such as prowess on the soccer field, can bring. And when faced with the prospects of financing the heavy expenditures of the *cargos* in comparison with the purchase of a radio, public address system to broadcast phonograph records, or even to purchase a truck, the latter is a much more interesting set of prospects.

An additional pressure to limit ceremonial expenditures in connection with the *cargo* system has come from the Church.[3]

3. An article from a Cuzco newspaper reported in 1971 that: "According to Dr. Oscar Velazco Lasteros, the head of the Parish of Anta Province, cargos will be prohibited after 1972 according to the new regulations of the Church; now, religious festivals will be celebrated in collaboration with the entire community. No one individual will bear the entire economic burden. In this way, prob-

A church decree prohibited individual *cargos* beginning in 1972 in order to reduce both the excessive individual expenditures and alcoholic consumption.

Another factor in the demise of the fiesta complex and *cargo* system is the loss of a direct rotation of religious *cargos* with political office. Here the main constraints lie in the relationship between the hacienda community and the independent community and the regional and national power structure. In Tukiwasi, as in other hacienda communities in the Pampa, before 1969–1970 there was no formal decision-making body through which resident peasants allocated power to a representative or set of representatives to organize an assembly and administer the community affairs internally and represent the community externally to the larger society. Instead, a hacienda administrator, the *mayordomo*, called together the *feudatarios* in a meeting and made the decisions as to how the time would be allocated over a period of two weeks. This "work plan" usually involved time spent during the week by *feudatarios* on the lands of the hacienda; weekends were set aside for work on the subsistence plots of the *feudatarios*. Any other activity of the community, such as road repair, maintenance of the church, or preparations for the annual fair, were organized by the *mayordomo* as part of the regular meetings. The position of the *mayordomo* was that of a delegate of the hacendado; his power to organize the life of the community and to represent the community to the extra-communal institutions stemmed from that role. His position was further reinforced by his religious activities; he taught catechism, held the *cargo* of *previste*, or keeper of the keys of the church, and continually reinforced Catholic ideology.

The anomaly that is interesting in this context is that Tukiwasi has an unusually strong religious *cargo* system and fiesta complex. The Church in the central plaza, constructed by one of the owners of the hacienda, is the home of the Señor de los Milagros, one of the three Señores, along with the Señor de Wanka in Urubamba, and the Señor de los Temblores in Cuzco, that form a triumvirate of three Señores believed to

---

lems due to alcoholic beverages will be avoided" (*El Comercio* [Cuzco] 28 December 1971).

have miraculous powers dating to their origin as part of the history of the Conquest of Cuzco. Each year during the fiesta of the Señor de los Milagros in Tukiwasi there is an annual fair that draws participants from all over the Cuzco Department. The fair lasts seven days and is one of the most important events in the Pampa de Anta. *Cargo* holders spent considerable sums in catering to fiesta participants. But, inasmuch as there is no internal organization of the community below the *mayordomo*, there are no political offices. What remains is the interesting fact that the wife of the major cargo holder, the *mayordomo* of the Señor de los Milagros, is allowed to carry a silver-encrusted staff much like the vara if not a vara; informants pointed out that it was not a vara, which would indicate authority, but merely a staff. This is consistent with the nature of the patron-client behavior prescribed by the hacienda structure.

In the independent communities of Rumipata and Antapampa the pattern is more complex. Due to the recognition of these communities under the provisions of the 1920 constitution, they are required to adopt a formal political organization as specified by law. A set of authorities is elected in open assembly, including a president, secretary, treasurer, and a *vocal* who deals with the internal affairs of the community, and a *personero* who represents the community to extracommunal institutions. When this formal political organization was authorized for the communities, the direct link between religious sponsorship and political office was considerably weakened. In practice, however, elected officials, especially the more important ones, came from the ranks of the elders who had held correspondingly important religious *cargos*. The past *personero*, for example, in Rumipata had held the now defunct *alferez cargo*.

The political organization as set out in the indigenous communities legislation was changed even further in the peasant communities law of 1970. This law established a dual organization of authority, an administrative council and a vigilance council, which worked in concert with the general assembly of each community. In this way, ideally a local system of checks and balances would maintain a viable political organization. The administrative council was the formal governing body of the community made up of a president, a vice-president, a secretary, a treasurer, and one or more *vocales*. The vigilance council is made up of a president, a secretary, and a *vocal*. To hold office in either the administrative council or the vigilance council a *comunero* has to be able to read and write.

While recognition as an indigenous community or peasant community is not the sole causal reason for the weakening of the civil religious cargo system, it did reinforce a trend toward a decisionmaking body based on the open assembly and the election of officers in that assembly rather than through a direct process of the successive sponsorship of religious cargos and political offices.

Political Organization: External

Another important aspect of the political organization of the communities is their links with the national juridic-political system. In Peru, administrative practice stems from the municipal law passed in 1873 that laid the basis for the present administrative infrastructure. The country was divided up into departments, provinces, and districts, each of which is an administrative unit with a capital. Above the district capital, or *aldea*, which is presided over by a *gobernador*, there is the provincial capital governed by a sub-prefect and the departmental capital governed by a prefect. Appointment to these positions is extremely centralized; the prefect and sub-prefect are appointed by the president of the republic while the *gobernador* is named by the prefect. The *teniente gobernador*, the formal link between the *aldea* and the peasant community, is named by the sub-prefect (Austin 1964:16–17). In peasant communities in the Pampa, a candidate for *teniente gobernador* is agreed upon in an assembly of the community and then submitted to the sub-prefect for final approval. Antapampa, Tukiwasi, and Rumipata each have a *teniente gobernador*. In Tukiwasi, the *teniente gobernador* is at the same time the representative of the hacienda, the *mayordomo*. This position is a further reinforcement of his power in the community.

The *teniente gobernador* is responsible for maintaining the legal apparatus of the national society on the local level. If any conflict or alleged violation of national law occurs in the community that involves arbitration according to the national legal code, then the *teniente gobernador* assumes responsibility. If legal action is required, he will usually initiate it through the lower channels of the juridical hierarchy. Likewise, if any juridical or political activity is involved with community members or the community at large, then the *teniente gobernador* will be the contact, or representative, of the national government at the local level. For example, if there is a conflict between communities as occurred during a land invasion of Antapampa land by members of Rumipata, then community members will go with the *teniente gobernador* to either the *Gobernador* at the district capital or most often the sub-prefect in the provincial capital. The *teniente gobernador* in such a case will present the problem to the top official.

Given the range of public officials that have some degree of power over the peasant, nevertheless the most important official is one most closely linked to the traditional political organization of Pampa communities. This official has been the *personero,* the role that most closely approximates local norms of authority. His importance is revealed in Figure 2–9,

Figure 2–9. Attitudes of Peasants toward Public Officials in a Sample of Haciendas and Independent Communities in the Pampa de Anta, 1964

Positions Considered Most Important	Independent Communities		Haciendas		Total	
	No.	%	No.	%	No.	%
Alcalde	59	14.9	13	12.5	72	14.4
Gobernador	84	21.3	10	9.6	94	18.8
Personero	214	54.0	36	34.6	250	50.0
Ten. Gobernador	18	4.5	—	—	18	3.6
Juez	7	1.8	—	—	7	1.4
Administrador/ Mayordomo	—	—	24	23.1	24	4.8
No Opinion	14	3.5	21	20.2	35	7.0
Total	396	100.0	104	100.0	500	100.0

Source: Table 54, Informe del Estudio Socio-Económico de las Comunidades 7 de los Fundos de la Provincia de Anta—Cuzco 1964, in files of the Ministerio de Agricultura, Cuzco.

which shows the results of a survey conducted in 1964 among a sample of haciendas and communities in the Pampa.[4] The *personero* was considered more important even in hacienda communities which by definition are not recognized and do not have an official *personero* position. What is not disclosed in this table is that, with the exception of the *personero*, all public positions are held by mestizos; in order to understand the implications of this fact, we will now turn to the relation between class and ethnicity in the region.

Class, Ethnicity, and Social Change

Another aspect of the social fabric of the communities is the nature and variation in horizontal class divisions in Rumipata, Antapampa, and Tukiwasi.

Until the last three decades, the social structure of the Pampa was characterized by a marked bipartite class division, at times approaching a caste division. On the one hand, there was a readily identifiable Indian population set off by a set of traits that included an Indian costume, reliance on the Quechua language, adherence to an Indian lifestyle based on

4. This survey was made with the object of declaring the Pampa de Anta an agrarian reform zone in 1965. The types and locations of the settlements selected for the survey are as follows:

District	Community	Hacienda	No. of Families	Number of Families Interviewed
Zurite	Ancahuasi		200	44
	Ccacajuara		96	14
	Curambamba		130	59
	Yanama		98	30
	Kataniray		140	12
	Tambo Real		120	40
		Chamancalla	60	15
		Ancachuro	50	12
Limatambo	Ayaviri		120	32
	Pampaconqa		250	30
	Chonta		280	56
		Sauceda	15	7
		Huerta Huayco	50	23
Mollepata	Mollepata		20	31
		Markahuasi	25	8
		Musamanpata	60	39
Chinchaypucyo	Chinchaypucyo		120	46

agriculture as a livelihood, and a core of nonspecific traits based on an indigenous cultural matrix. While there was some movement between city and countryside, Indians preferred to identify with the attitudes and values that characterized their community of residence. The mestizo class, on the other hand, adopted a lifestyle and set of attitudes and values that corresponded to the sierra urban subculture of Cuzco, the departmental capital. Mestizos resided primarily in the lower-ranked urban localities of the Pampa: the provincial capital of Anta, and the district capitals of Anta, Huarocondo, and Zurite. They were also to be found in the peasant communities, where they occupied roles such as shopkeeper, merchant, hacienda administrator, *pequeño propietario,* and other similar occupations. The livelihood of the mestizo class in the Pampa depended on the acquisition of the surplus generated by the peasant agricultural sector of the regional economy.

Local terms used to describe social groupings in the Pampa reflect this bipartite class structure. From the point of view of the mestizo, the indigenous residents of communities and haciendas are Indians. From the perspective of the peasant living in the rural community, two terms are used: *misti* refers to mestizo residents of the communities and the town dwellers. *Runa* is reserved for residents of a peasant community who display indigenous traits, participate in public fiestas, and concern themselves with the internal affairs of the community. In their world view the local community is the parameter for day-to-day social interaction. Extra-communal institutions are perceived by *runas* as essentially distinct. They are not necessarily perceived as hostile but, at the least, as qualitatively different from social life in the community.

The use of *runa* and *misti* by peasants indicates a traditional correlation between ethnicity and class in the Pampa de Anta. During the last three decades, though, the distinction has been challenged by the effects of social change and acculturation. It is now difficult to distinguish Indians in terms of "traditional" indicators of language and dress. Dress is no longer based on a homespun folk Indian costume, except in the case of a very few older women in each community, but now approximates what might be referred to as a hybrid "cholo" costume characteristic, with some exceptions, of highland acculturated peasant groups throughout Peru. The Quechua language, as an important indicator of "Indianness," continues to be a necessity for communication among the peasant population but a growing number of peasants, men in particular, are quite at ease in Spanish conversation. Nevertheless, in comparison with the highly visible "Indian" population of Quispicanchis or Paucartambo, for example, the peasant population of the Pampa de Anta is much more acculturated.

There are a number of processes, including education, marketing, and military service, which are responsible for increasing the frequency of contacts between the community and the larger society. One of the most important is the effect of migration processes, in particular seasonal and return migration. Seasonal migration is the temporary migration of an individual to either Cuzco, where he may work as a mason, *albañil,* a burden carrier, *cargador,* or a construction helper, or to a location in the tropical valleys, such as the Convención or Tambopata, where he will work on the harvest of coca, tea, or cacao. This type of migration occurs during the off-season of the agricultural cycle, from June to August. The dead season in the Pampa blends in perfectly with the peak harvest season in the tropical valleys. Seasonal migration to the valleys has by now become quite institutionalized with a system of labor recruitment in Izcuchaca and in Cuzco where peasants are contracted and transported to their destinations. Return migration, on the other hand, is the decision to return to the rural community following a residence outside, usually in an urban setting. It is the reverse "feedback" from the pattern of out-migration alluded to earlier in this chapter.

Return migration is more of an urbanizing experience, since most peasants return from a stay in Cuzco or Lima, than seasonal migration. Informants state that the knowledge that one is returning to his hearth and home in Anta considerably tempers the attitudes and values that are adopted during the short work period characteristic of seasonal migration.

Figures 2–10 and 2–11 provide valuable insight into the flow and frequencies of return and seasonal migration, which are based on the migration sample.

Figure 2–10 reveals that approximately half of the peasants interviewed had had some kind of migration experience outside their communities of residence. Figure 2–11 presents the locations to which peasants reported having migrated. Sea-

Figure 2–10. Peasants Having either Permanent or Seasonal Migration Experience as a Percentage of the Total Sample

	Migration Experience	Percent
Rumipata	27	58.7
Antapampa	12	41.4
Tukiwasi	7	53.8
Total Sample = 88	46	52.3

Source: Migration History Sample.

Figure 2–11. Places of Departure Reported by Migrants

	Permanent				Seasonal	
	Cuzco	Lima	Valleys	Other	Cuzco	Valleys
Rumipata	5	4	3	1	8	13
Antapampa	1	1	1	—	5	8
Tukiwasi	3	1	2	—	1	3
Total = 46	9	6	6	1	14	24

Source: Migration History Sample.

sonal migration, especially to the subtropical valleys, is the most common type. Responses from Tukiwasinos indicate that seasonal migration is as common there as in the independent communities. Inasmuch as the hacienda economy of the Pampa is tied to the seasonal agricultural cycle, *feudatarios* are as free as *comuneros* to leave the community during the dead season.

Besides seasonal migration, return migration from a permanent residence in either an urban setting such as Cuzco or Lima or the subtropical valleys is a significant pattern. While Cuzco is enumerated as a more frequent place of destiny than Lima, as has been pointed out earlier, most recently Lima has offset Cuzco as appealing to migrants. This change is apparent by comparing the migration histories of permanent migration presented in Figure 2–11 with the location of offspring of peasants in Figure 2–8.

The incidence of return migration supports the statements of informants that migration to the city is heavily risk laden. Since out-migration has become frequent, there has been a constant but low rate of return migrants who for whatever reason did not remain in the city to live. These individuals are quite conscious of their perceived socio-cultural differences; upon return they initially attempt to set themselves off from the residents by continuing to dress in city clothes, objecting to speaking Quechua, and expressing a preference for urban tastes in cigarettes, food, and liquor. Nevertheless, return migrants and immigrants alike were faced with strong social controls to conform to locally prescribed codes of behavior. Gossip, threats of witchcraft, and envy, to name a few such mechanisms, continued to be effective in this regard. Most individuals acquiesced; the only alternatives were complete isolation from social interaction in the community or interaction within the regional mestizo sphere. Mestizo immigrants, usually occupationally specialized and linked to the regional economy of the mestizo towns, preferred the latter alternative.

Figure 2–12, derived from the Peasant Communities Census, provides an insight into other important indicators of social change in the three communities. These figures suggest the importance of education in social change; the degree of bilingualism (as measured by Spanish speakers) is strongly correlated with having one or more years of education. Possession of a voting certificate or military document is a response to a question asking for an identification document of some kind.

The result of these and other processes of social change and acculturation is a change in the perception of ethnicity from the point of view of both the peasantry and the mestizo classes. In the former case, peasants will now extend the term *misti* to include those who have manifested a willingness and ability

Figure 2–12. Selected Indicators of Social Change in Rumipata, Antapampa, and Tukiwasi

	Enumerated Household Heads	ing Spanish Speak-	%	One or More Years of Education	%	Voting/ Military Certificate	%
Rumipata	218	109	50.0	110	55.0[1]	49	25.5
Antapampa	143	59	41.3	59	41.3	40	30.0
Tukiwasi	53	19	35.8	21	39.6	24	45.3

1. Education data is missing for 18 household heads; therefore, the base number is 200 rather than 218 (household heads). Source: Peru, Ministerio de Trabajo, 1970

to deal with macro-societal institutions on its terms. Peasants will characterize such an individual as "one who has familiarity," or "*uno que tiene roce.*" In order to merit this distinction, a peasant must demonstrate to his peers in the community that he has successfully dealt with a significant challenge outside the community in the larger society. For example, one who is born in the community but later migrates to Cuzco or Lima, establishes residence there, and leaves community life for some time, upon his return is regarded as a *misti,* even though he may take up subsistence agriculture and by all other indicators re-create his lifestyle along peasant lines. Another indicator is having held a political office, particularly the position of *teniente gobernador,* that requires dealing with mestizos on a regular basis. Most importantly, though, the expanded *misti* perspective implies an attitudinal change that indicates the willingness to deal with members of the governmental bureaucracies and other mestizo officeholders on their terms.

A second indicator that social change has resulted in an acculturation of the peasant population is the use of the term *cholo* by Cuzco townspeople to refer to the peasants residing in the Pampa. This term implies a recognition that Pampa peasants are no longer *indios,* as, for example, they would characterize the Indian population of Quispichanchis or Paucartambo, but rather *cholo,* a term that signifies an aggressive acculturated peasant. There is a clear sense of fear by mestizos of the *cholo* as an unknown, a psychological realization of the dominant status encountered by peasants under the aegis of mestizo superordination.

While social change and acculturation have occurred during recent decades it is important to realize that the basic bipartite class division remains essentially the same. The perseverance of the *misti-runa* terms of reference, although changed in their ethnic content, continues to reflect a class division reinforced by the political economy of the region.

The Political Economy of the Pampa

The base of the political economy of the Pampa until 1969 was power allocated to representatives of the mestizo class by an hacendado elite operating out of the departmental capital of Cuzco. Control over land was the only economic resource available in the economy of the southern sierra that was controlled by residents of the southern sierra. While other non-agricultural resources existed, such as the mining, textiles, tourist, beer, and fertilizer industries, they were either under-capitalized and inefficient, or more commonly relegated to the status of an economic enclave such as tourism controlled by coastal interests and channeling profits out of the sierra. The basic economic resource in the southern sierra has traditionally been land; those who control access to land have been able to manipulate economic and political institutions independent of constraints from the coastal center.

Although mestizos are quite occupationally diversified and live in both the *aldeas* and the peasant communities, their economic livelihood depended on in one form or another on the appropriation of economic resources of the peasantry in the countryside. There were a number of ways in which mestizos appropriated these resources. The hacienda administrators, *mayordomos,* were essentially middlemen in the expropriation of the funds of rent exacted from *feudatarios;* their livelihood depended on the persistence of the hacienda as an economic unit and the continued support of the hacienda elite in the maintenance of a political economy biased in favor of mestizo interests. At the upper levels of the hacienda structure, that is, the hacendado elite, which was usually located in the departmental capital, the livelihood and lifestyle depended on the surplus generated by the funds of rent of the peasants. Shopkeepers and other mestizos residing in the rural communities depended upon the acquisition of produce at favorable terms in exchange for cash or kind. The entire "urban" economy of the lower-rank *aldeas* depended on a favorable exchange rate in the acquisition of the marketable surplus brought to the towns by the peasants, and the *pequeños propietarios* depended on minimal wages paid to landless and landhungry peasants to make their own enterprises efficient while they maintained residences in Cuzco. Much of their cultivation requires access to tractors and other mechanized implements that were obtained from haciendas and townspeople at rates that were exorbitant for peasant cultivators.

The relations between peasants residing in communities and the varying administrative levels were based on the complete domination by governmental officials residing in the lower-ranking *aldeas* and provincial capitals. Domination occurred in a number of forms. The sub-prefect, in practice, has been the major spokesman for administrative policy of the government; all major decisions in the region are usually made by the sub-prefect. There has been a clear policy of laissez-faire toward the activities of the hacendado elite. This is a reflection of the bias of the prefect in the departmental in favor of the hacendado elite that is passed on to the sub-prefect in the provincial capital. Over the years, alienation of communal land, labor abuses involving *feudatarios*, boundary disputes, and other conflicts involving peasants and adjoining haciendas have invariably been decided in favor of the hacendado. Numerous exploitative practices, of which the *muyuy* is most symbolic, have gone unchecked by a policy of hands-off by the government officials. Until the practice subsided recently there was a tradition, noted by Sabogal Wiesse (1969:175) called *muyuy*, in which each community was obliged to deliver animals on occasion to the sub-Prefect, the comandante of the Guardia Civil post, and the district *Gobernador*. Although this practice had disappeared, it remains a tradition that represents a considerable sore spot in the sides of the *comuneros*. During 1972 there was an open charge made against the then acting *teniente gobernador* in Rumipata that he had obliged a *comunero* to deliver a sheep to him. Although the charge later proved to be false, it was the cause of considerable verbal abuse in the assembly during that period.

If any conflict is brought to the judicial system in the form of a *juicio*, traditionally it ended up unresolved because of bureaucratic red tape. Litigation involves numerous transactions made on official stationery, *papel sellado*. Each transaction may cost the equivalent of a day's wages. If the litigation reached the resolution stage, it was almost always decided in favor of hacendado interests. An independent judiciary that would hear cases involving peasant suits in an objective manner virtually did not exist.

Perhaps the major source of power at the disposition of the sub-prefect is the Guardia Civil post located in Izcuchaca, which could be used in situations involving any major conflict with peasants.[5] During the early 1960s, peasants from Rumipata invaded the lands of *pequeños propietarios* in Antapampa. The Sub-prefect resolved the situation by sending a contin-

5. For a discussion of the role of the Guardia Civil in the political structure of the Department of Cuzco see Pierre L. van den Berghe and George P. Primov (1977:73–74).

gent of the Guardia Civil along with him to the community. Although the sub-prefect attempted to regain possession of the land for the *pequeños propietarios* verbally, his access to the Guardia Civil was an important negotiating resource. In this case, the land held by the *pequeños propietarios* was claimed by the Rumipata peasants to belong to their community and to have been alienated from community control during an earlier period.

Throughout the early 1960s massive repression by the Guardia Civil was a common response to the widespread wave of land invasions in the Pampa de Anta and in effect throughout the southern sierra. One of the largest haciendas, Huaypo Chico, in the Pampa was invaded by peasants in 1963 from the neighboring community of Chacan. Later in the year, around two thousand peasants surrounded Anta and demanded that the Huarocondo hacienda be turned over to them. The next year a series of invasions occurred on the haciendas Andenes, Campana Oroco, Kenko, Santa Bárbara, and La Perla of Zurite district, and Pitucalla, Kallanquiray, and Larraconca of Huarocondo district (Roberto Mac-lean y Estenos 1965: 134–45).

Following a split in the leadership of the peasant movement in the Convención Valley, with which the peasant movement throughout the southern sierra was intimately connected, and the widespread use of force, peasant mobilization in the Pampa subsided, and the pattern of mestizo control over crucial institutions was able to be re-created (Wesley W. Craig 1968; Anibal Quijano Obregon 1965). The ability of the hacendado elite in Cuzco to activate its institutional support in the Pampa is a reflection of its continued strength during this period. The power structure survived the peasant movement of the early 1960s, and the movement has never recovered on an organizational level equivalent to that period.

As a result of the invasions in the Pampa there have been more subtle changes. As Susan C. Bourque (1971) has noted in other parts of Peru, there have been subtle attitudinal changes in the peasant population as a result of participation in organized peasant mobilization. Peasants are more sophisticated about political activity, aware of the unstable character of the political economy, and less accepting of subservient behavior such as the *muyuy*. More importantly, out of the land invasions of the 1960s there emerged a "peasant leader" role that, as will be seen, provided a niche in the political economy that would be manipulated by political entrepreneurs in the period following the 1969 agrarian reform.

For the most part, members of the mestizo stratum of the Pampa were not directly involved in the exercise of power; they preferred to deal with their representative in the juridic-political institutions when the occasion arose. Nor did they participate in the internal affairs of the peasant communities; those *pequeños propietarios*, shopkeepers, teachers, and merchants who lived in the countryside did not ordinarily interact with peasants above a minimal level necessary to recruit labor or maintain a clientele. Because of the dispersion of mestizos throughout the countryside, in one sense they were "atomized." They did, however, participate in a mestizo sphere of interaction.

Such interaction among mestizos tended to occur in Izcuchaca. All provincial business and most district business for Anta district is carried on in Izcuchaca; there the sub-prefect has his home and provincial offices, and all major governmental offices are located there. More importantly, Izcuchaca-Anta is the gathering place for the mestizo stratum of the countryside. It enables mestizos to socialize in any of a number of contexts: in the market, on the Cuzco-Izcuchaca bus lines, and most often in the bars, *cantinas*, that serve beer and cater to an exclusively mestizo clientele. These bars are distinct from the *chicherías* that serve *chicha* (an indigenous beer), and *trago*, rum, and are frequented by an exclusively peasant clientele. In these instances relations are made and maintained among all of the different groups making up the mestizo stratum: the *pequeños propietarios* of the countryside, other mestizos living in peasant communities, hacienda administrators, teachers, and the townspeople including the members of the different government bureaucracies including the sub-prefect, the Guardia Civil, and others. These relations that are created are often translated into more enduring bonds based on friendship, marriage, and often *compadrazgo* ties. For example, virtually every *pequeño propietario* can trace both real and fictive kin ties with the major hacendados of the region. These ties are often created through intermarriage or through adoption of a trusted individual into the immediate family of the hacendado. They may also result from social interaction of an informal nature, and recognition of mutual interests vis-à-vis the peasantry. The solidarity that obtains from such ties is transformed into networks of contacts that effectively bind together the different groupings of the mestizo stratum into a class grouping that transcends, in many cases, the localized residence of the mestizos living in the countryside.

Information Flow

The mechanisms through which information is disseminated concerning activities originating in the macro-society will now be examined, in particular those activities such as national policy affecting peasants. The usual path is a combination of those governmental institutions that maintain links to peasant communities and the media. I have mentioned earlier the manner in which the rural community is connected to the juridic-political network through the *teniente gobernador*. In addition, recognized communities, such as Rumipata and Antapampa, have an internal political organization including a specific political office, the *personero*, charged with representing the community to the office of peasant affairs located in Cuzco. Information regarding national policy that implements developmental strategies involving peasants is disseminated by representatives of governmental organizations, such as the Ministry of Agriculture; the cooperative agency; the agrarian reform agency; and the community development agency; which operate out of offices in Cuzco, or at the provincial level in Anta-Izcuchaca. The media consists of two main sources: the newspapers, such as *El Sol* and *El Comercio* of Cuzco, and the radio stations operating out of Cuzco. Periodically there had been a column in *El Comercio* reporting provincial news; some information about activities in Anta comes to light from this source. The major content of the radio programs has been highland folk music, but some news concerning regional and national events is also announced.

Ordinarily news concerning a developmental policy originating at the national level would be disseminated through the bureaucratic channels of the particular agency involved. The community in turn would obtain information through a visit by representatives of the agency. Depending on the scope and the impact of the policy, the juridic-political network and the media would also be activated. The following diagram illustrates these "formal" sources of information.

Figure 2–13. Formal Information Flow

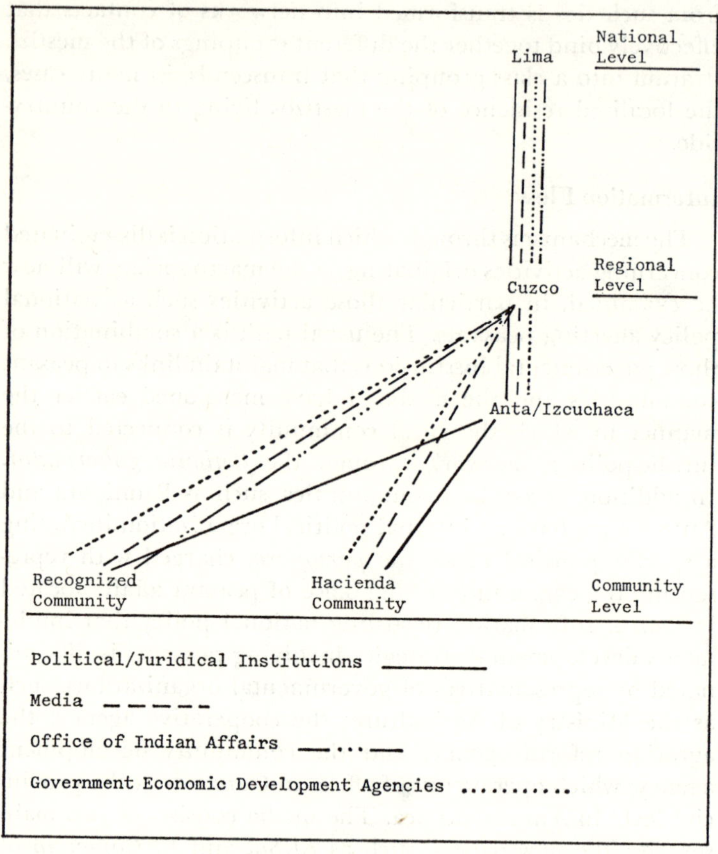

Other more informal sources supplement the formal information network and provide much of the information upon which peasants base their perception of the activities. These sources are those individuals who have occasion to participate on a periodic or regular basis in activities outside their community. While such activity may not directly involve the procurement of information as an end, it does bring these individuals into contact with other sources of information outside the community.

Many individuals are involved, either directly or indirectly, in the transfer of economic surpluses, from peasant households to the larger society. The hacienda administrator was instrumental in the appropriation of the labor and produce derived from the funds of rent exacted from *feudatarios* residing in hacienda communities such as Tukiwasi. Other activity included the transactions carried out by middlemen in the transfer of the agricultural surplus from peasant households to extra-communal markets. And seasonal migration for the purpose of selling one's labor during the agricultural dead season is another example of interaction in the larger society deriving from the transfer of economic surpluses.

Another mechanism through which peasants interacted with extra-communal institutions was through the administrative network of the Catholic Church. The locus of this interaction is Anta, the provincial and district capital; the provincial religious offices are located there, and any activities such as confirmation, baptism, and other formal religious observances take place there. Outside of movement to Anta for these ends, there was little activity carried out by priests in the community, in particular Rumipata and Antapampa. Tukiwasi is more fully integrated into the administrative network because of the important position of its church in the departmental religious calendar and through activities of the religious instruction program.

Another periodic source of information came from the visits of friends of relatives who had left the community to reside in an urban center where they had the opportunity to be exposed to the media and exchange information with other residents of the city. These visits are infrequent and occur mostly during the Patron Saint Carnival festivities in each community. And lastly, a major source of information comes from contact with mestizos living permanently in the communities either as *pequeños propietarios* or as shopkeepers, merchants, or other occupations based on a peasant clientele. As I have mentioned, mestizos in these situations did not participate in the political activities of the communities but were involved in economic transactions and occasional social intercourse. Figure 2–14 illustrates these kinds of informal sources of information in contrast to the formal communications network presented in Figure 2–13.

There is considerable variation in the extent to which these sources, both formal and informal, provide culturally relevant information to peasants. First, numerous problems exist with respect to the transmission of information within formal channels. To use either the newspapers or the radio as

Figure 2–14. Informal Information Flow

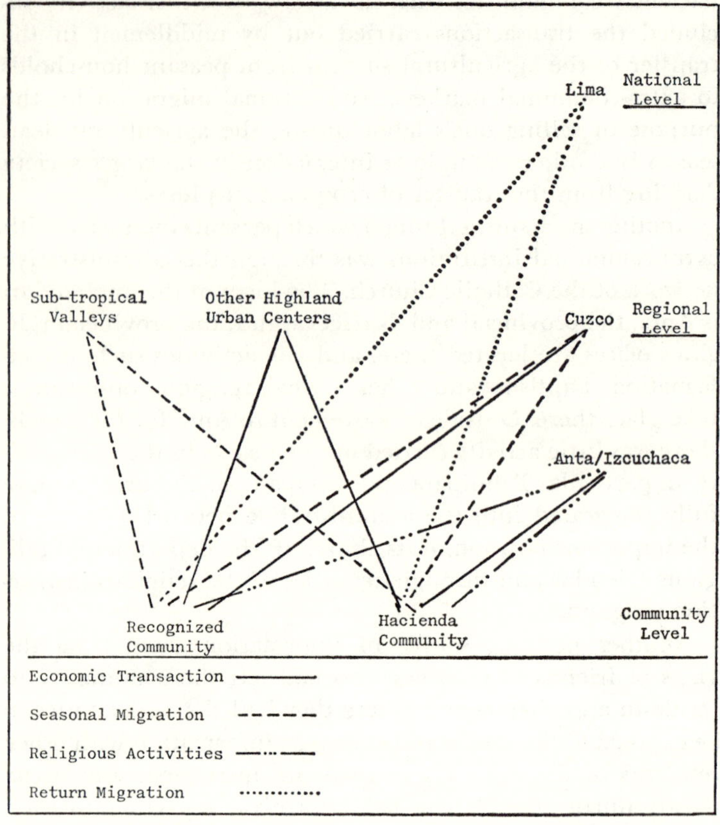

a source, one must be fluent in Spanish; in the former instance, newspapers require a rather sophisticated reading ability. There are no regular news programs in Quechua on the radio.

In practice, formal media channels have a negligible effect on peasants. The few newspapers that manage to enter the community are read almost exclusively by the more literate members, while the majority of the peasants prefer to listen to the radio as a source of highland *huaynos,* folk songs, rather than as a source for any other kind of information.

Most of the governmental agencies dealing on a regular basis with peasants are severely handicapped by a lack of personnel and a mobile and flexible bureaucracy that can disseminate information on a stable basis. A preference for employees to stay in the urban centers, a lack of transportation, and a number of other problems combine effectively to constrain the visits to the communities required to carry out the role of the agency as information source. In place of these visits, peasants must come from the countryside to the offices of the agencies, either in the provincial capitals, or more often in the department capital, Cuzco. These trips mean time must be set aside from agricultural chores, money expended to pay transportation costs and, often, lodging while in Cuzco.

These limitations would seem to be inherent in most programs of cross-cultural information diffusion. However, there is an additional important element in the case under study here. In the southern sierra, virtually all of the important positions through which information flows, including the teniente gobernador, representatives of government bureaucracies, teachers, priests, and hacienda administrators are members of the mestizo stratum of the regional political economy, and from the perspective of the peasants in communities, *mistis*. In effect, control over sources of information has been a manner in which the superordinate class position of the mestizo grouping could be maintained and thus support additional power delegated to lower level members of the mestizo stratum by the hacendado elite operating at the level of the departmental capital.[6] Thus, while situated in a superordinate class position with respect to the peasants in the Pampa de Anta, the mestizo stratum has also acted as an intermediary in the flow of information from the national society to the peasant community.

6. This is a general occurrence throughout the Peruvian sierra. As Cotler states: "The necessary and sufficient requirements of the system of domination described are a function of the mestizo's access to the system of authority through his knowledge of Spanish, and through literacy which allows him to elect or be elected to and to designate or be designated to fill positions within the system of national authority or within the public administration. . . . The mestizo is not only the owner or administrator of the hacienda; he is also lawyer, middleman, judge, governor, policeman, trader" (1970:420).

3

The Organization of Production in Pampa Peasant Communities

In this chapter we will focus on the peasant economy of the Pampa de Anta, in particular, looking at the organization of production. We will draw heavily on field work in the three communities, but where possible, attempt to generalize to the level of the region. In the first part of the chapter, the household sphere of the economy will be analyzed; it will be followed by the supra-household. There are significant differences at the supra-household level between the independent and the hacienda community; these will be noted in the course of the chapter.

As can be ascertained from Figure 3–1, the majority of the inhabitants of Rumipata, Antapampa, and Tukiwasi are farmers. The salaried employee category refers to individuals who work in Izcuchaca or locally as employees of government agencies or for private business enterprises; laborers, on the other hand, usually sell their labor on a day-to-day basis to mestizo employers or to public works projects, particularly road construction. Some craft specialization such as weaving, carpentry, and woodcutting, can be found, particularly in Antapampa, but it contributes little to the overall picture of the economy. Most small craftsmen or shopkeepers will either directly or indirectly cultivate a subsistence plot that provides a substantial portion of the household income.

Figure 3–1. Primary Occupation of Household Heads

	Farmer	Salaried Employee	Laborer	Other
Rumipata	193	3	6	13
Antapampa	112	4	3	24
Tukiwasi	48	—	1	4

Source: Peru, Ministerio de Trabajo, 1970.

The basic unit of the economy is the household. The household head allocates the resources of land and labor at his disposal to meet the subsistence needs of his household. Any surplus that remains after subsistence needs are met will be sold on the market using middlemen or directly in the marketplace in Cuzco.

Resource Distribution and Characteristics

Land is the most important resource of the household. Without access to land the household would not continue in existence. Yet even with secure access to a suitable quantity of land there are a number of constraints that reduce the prospects for a subsistence return from household labor. These constraints are continually weighed by the household head in the allocation of resources at his disposal in meeting subsistence.

In the ecological setting of the Andes, one of the basic principles of subsistence has been the acquisition of the products of vertically arranged ecological zones. This principle was first discovered by the ethnohistorical investigations of John Murra (1972) and has since been found to permeate the economic and ecological patterns of numerous rural communities, particularly in the central Andes where the influence of altitude is so great (Steven Webster 1971; Daniel Gade 1975; Stephen Brush 1977; Dean Arnold 1975; Benjamin A. Orlove 1977). Each of the communities is located at the edge of the flat bottomland of the Pampa and thus in a position to control land on the floor and on the slopes of the surrounding mountains. These two natural ecological zones circumscribe very different agricultural regimes.

The first regime, referred to locally as the pampa, is the land on the floor of the valley, lying at about 10,180 feet in altitude. It is provided with a permanent source of water from small streams, *riachuelos,* running down the mountain slopes, and also from natural springs originating in the communities. These streams eventually empty into a river-lagoon system in the interior part of the pampa. Each community having access to land in the pampa can build and maintain a permanent irrigation system based on the channeling of the hydraulic output of these streams. Communities vary in the extent to which they exploit this resource, though. Antapampa, for example, has a clean and well-maintained irrigation system in which households are allotted a quantity of water per month and pay taxes for this right to a *juez de aguas,* or water official, elected by the community. Rumipata, on the other hand, lacks a specific infrastructure to regulate irrigation, and individual use is based on individual responsibility.

Although water is available on a permanent basis in the pampa, it is difficult to control efficiently. Soil does not drain well, and during the rainy season, standing water often accumulates in the pampa fields and soon stagnates, causing root rot and disease. Potatoes are particularly susceptible, while maize is more resistant. For this reason, and the fact that maize does not grow well on the mountain slopes, the pampa is principally a maize-producing zone. Potatoes and some other crops, such as the vetches, *habas,* and *arbejas,* which are interplanted with maize, are also grown, but infrequently. The major exception, the recent introduction of a cash-cropping sequence of an early potato/onion cycle, will be discussed later. Agriculture in the pampa is constrained by the uncertainties of frost, hail, and other natural hazards but to a considerably less extent than in the higher slopes of the puna.

The puna is the second ecological zone, ranging from the floor of the pampa at 10,180 feet to the upper limits of the cultivable reaches of the mountains at around 11,000 to 11,500 feet. Because of erosion and disadvantageous topographical features not all of the land is of the same quality. Each microvariation in altitude or location with respect to topography can have a considerable effect on the potential of the cultivable land. Rumipata, for example, is located in a ravine protecting its lands from extremes in temperature variation and harsh wind found in the more exposed communities such as Antapampa and Tukiwasi. As a general rule, though, the puna is extremely susceptible to hail and frost, and the higher the altitude the greater the risk of these calamities.

To ensure a sufficient level of fertility in puna soil, a sectorial fallowing cycle (Wolf 1966:56) is followed. Puna land is divided up by the community into sectors, called *tiray,* according to the total number of years in a cycle. Rumipata has seven sectors for its seven-year cycle, while Tukiwasi has two for a two-year cycle. A cycle involves a series of crops cultivated in sequence during a growing period followed by a fallow period. The length of the fallow cycle is an important index of the quality of the land; length is set by the quality and amount of land available. Rumipata, with a four-year fallow period followed by a three-year growing period, is much better off than Tukiwasi with a one-year, back-to-back growing/fallow cycle, especially in view of the more protected position of Rumipata's lands.

The puna zone is the principal producer of potatoes and other indigenous rootcrops, such as *oca* and *papa lisa,* the vetches, *habas* and *arbejas,* and indigenous cereal substitute *quinoa.* Wheat and barley are grown also. Cultivation practices are determined by the amount of land an individual has in a particular sector, and the rules with respect to the crops called for by the rotation cycle. Preferences for crops are usually based on mixing a sufficient quantity of potatoes and *habas* with a proportion of other grains and tubers. Such a mix will reflect the tastes of each household. As was mentioned earlier, the other staple, maize, is grown in the pampa.

Another aspect of the distinction between pampa and puna is the variation they present in the ability to provide forage for animal husbandry. Most households maintain a small herd numbering from one to two mature cows or oxen to as many as eight to ten. These animals provide a supply, although minimal, of dairy products, such as milk and cheese, which when sold to middlemen locally or in Cuzco is a small addition to the cash reserves of the household. Oxen, in addition, are used in plowing the owner's plot, rented out, or used in reciprocal exchanges. But, most importantly, cattle are a form of instant liquidable savings that can be rapidly transformed into ceremonial expenditures encountered upon assuming a fiesta *cargo,* or bearing the costs of funeral or wedding expenses. Other unforeseen and major expenses such as litigation will be met by selling a cow. Maintaining a small herd, then, offers a minimal if constant return over the life expectancy of the animal and the instant liquidity of sale.

There is considerable variation in the quality of forage in the pampa and the puna. Generally, the pampa provides pasture of good quality and, despite the threat of intestinal parasites and disease in the humid and poorly drained bottomland, animals grow rapidly and maintain a high yield of dairy products. But pampa land is also prime agricultural land and thus competes with its use for forage.

Pasture is of lesser quality in the puna. The only exception is the few months of the rainy season where there is lush and ample high-quality pasture for all. With the end of the rains and the onslaught of the dry season, grasses dry up in the puna and the quality of pasture is poor. Foraging does not compete with agricultural use of puna land, though, as there are numerous sectors in the fallow stage of the cycle. In these sectors, grazing is an efficient means of returning manure to the soil. Pasture is also available above the timberline. The major constraint in the upper reaches, though, is distance. The distance factor means that either a person from one's own household must herd the animals full time, or indirect arrangements are made with other households residing in the puna. These arrangements are often quite risky. An advantage, to pasture in the puna at any rate, is that the incidence of intestinal parasites and other diseases decreases.

Reference has already been made in Figure 2–4 to the distribution of land in the three communities by ecological zone. However, the actual land that is farmed by any given household is a function of two factors: ownership and usufruct.

Ownership corresponds to an individual or group's claim to land that is legitimized through recognition by extra-communal legal institutions. Articles 208 and 209 of the Constitution guarantee that land within the boundaries of Indian communities is de jure owned in common by the community. This is based on an assumption that the roots of the contemporary peasant community lie in the pre-Incaic corporate landowning residential unit, the *ayllu,* that was subsequently incorporated with some modifications into the Inca state and that has undergone numerous modifications since the Conquest.

The ability of the communities to defend their boundaries against outside encroachment and internal dissension has weakened considerably during the period since the Conquest. Although this process has occurred throughout the Andean region and its elements are extremely complex, in the case of the communities in the Pampa de Anta, some factors can be singled out: (1) a series of demographic changes including a sharp decrease in population following the Conquest and a gradual increase during the last two centuries caused considerable shifts in the equilibrium that obtained between population and resources of the communities; (2) a more recent rapid increase in population, documented in the last chapter, due primarily to a reduction in death rates and the stabilization of birth rates; and (3) the diffusion of market forces during the colonial and later periods that resulted in the commoditization of Indian lands and their subsequent alienation from community control.

A direct result of these processes, in particular the latter, has been the rise of the hacienda system as a major landholding unit in the Pampa. This rise has been accomplished by manipulation of macro-societal legal/political institutions in order to legitimize ownership of land taken from Indian communities. Although a thorough analysis of the process in the Pampa de Anta is outside the bounds of this study, it would appear that the evolution of the hacienda peaked during the latter part of the nineteenth century and was revived during the 1930s when Ezequiel Luna, the most important landholder in the Pampa at the time, was elected senator during President Leguia's regime. Luna used his administrators to seize the property of adjoining peasant communities and to expand his own hacienda boundaries. As a result, some of the communities reacted strongly. In Rumipata, following a long struggle between the *personero* and the representatives of Luna, a decree from the president guaranteed the inalienability of the communal lands. Except for Luna's interlude, it appears that the hacienda began to decline in importance during this century and to fractionalize through inheritance patterns.

Tukiwasi is a good example of fractionalization similar to that which occurred in other haciendas in the Pampa. Following the death in 1910 of the owner the property was split up among six sons, each with one share, *acción*. Three sons sold their shares to an individual who was related in no way to the original family. He died in 1946, leaving the property to his sister. She died the same year and left the property to a relative; his widow now owns it. One of the original sons inheriting the property in 1910 bought the share of one of his brothers. He died in 1934, leaving the property to his wife and son. They in turn sold the property in 1941 to the present owner. The remaining share from the six of the original transfer was split up in eight parts in the will of the brother who died in 1951. The heirs decided to sell their minuscule plots to the present owner in 1953. At present, Tukiwasi, which until 1910 was one hacienda under one owner, at the time of the agrarian reform was in effect split up into three separate "haciendas" each with a separate owner. The *mayordomo,* however, administers the set of properties as one hacienda.

The communities of *feudatarios* that grew up within the boundaries of haciendas, such as Tukiwasi, were a response to the need for a resident supply of labor. Property in the hacienda system was owned de jure by the hacienda although originally traceable to Indian lands, because the regional political economy and its legal institutions were subservient to the interests of the hacienda sector.

While de jure ownership refers to an official recognition of an individual or group claim to possess rights to land, there are a series of gradations in the degree to which ownership manifests itself in practice. This refers to de facto ownership patterns. There are two basic types of de facto ownership that describe very different exploitation patterns. The first, in effect, groups hacienda property and communal property together in the sense that both represent corporate landholding units, recognized de jure, which are at the same time organized internally into units of production. Their internal organization includes, further, a series of types of "property" as a result of the privatization of land and the diffusion of market forces.

In the hacienda, resident *feudatarios* exchange their labor for access to plots lying in hacienda property in the pampa and in the puna. In the pampa a *feudatario* was given a certain plot as his *individual* property to be retained as long as he fulfilled his obligations to the hacienda. In the puna, land was set aside in a sectorial fallowing system as has been described. Property likewise was assigned to individuals to exploit. The bulk of the hacienda properties, though, was administered by a representative of the haciendas, who made decisions as to the allocation of labor provided by the resident *feudatarios.* These decisions were made in weekly meetings held by the representative and the assembled *feudatarios;* the weekly chores on hacienda lands were set up, and time, usually on Saturdays and Sundays, was set aside for them to cultivate their own plots. In Tukiwasi, individual *feudatarios* did not have a peculiar claim to the individual plot of land given them by the hacienda; the hacienda owner or his representative maintained absolute control over all land granted to *feudatarios.*

In the "free" communities of Antapampa and Rumipata, three types of de facto ownership can be found. The first refers to each *comunero*'s right to pasture and other resources in "communal" lands located in the high puna above the highest point at which land could be cultivated. By acceptance into the community an individual automatically gained rights to resources located in this zone that are held in common and not subject to individualization.

In the cultivable reaches of the puna zone, land is distributed, ideally in equal quantities, among all the community members. Cultivation of a plot or series of plots over time has resulted in the recognition by members of the community that it is the individual's property to be passed on to his heirs. This land may not be bought or sold, though, on the open market and at least theoretically reverts back to the community should it not be cultivated. To maintain an equal distribution, communities ordinarily redistribute land that has reverted back to the community through abandonment or death of a *comunero* leaving no apparent heir. Such redistribution is usually performed by the *personero* at an annual meeting and visit to the puna lands.

In practice, because of the failure of land to be returned to the community, considerable land accumulates in the hands of individual households. In Rumipata, for example, some individuals lay claim to as many as forty separate plots in the puna but lack of manpower often results in only a small percentage of these plots being cultivated.

In the pampa another pattern of ownership is apparent. Here land is held as private (as distinguished from individual) property: de facto ownership of property that may be bought or sold. While the distribution of holdings in the pampa is more egalitarian than in the puna, privatization of pampa land has resulted in some plots being alienated from community control by sale to outsiders. Virtually all pampa land is cultivated even when an "owner" is absent from the community.

A further aspect of the distribution of land is the use of a plot of land as distinguished from ownership. Ownership of land quite obviously implies the right to cultivate it; this is a right that is not always exercised, though, especially in the puna zone, where some households possess considerable amounts of land that go uncultivated.

There are a series of arrangements through which households with no land or with insufficient land can get access to excess land of other households. These involve indirect exploitation of land as contrasted with direct exploitation of land in which the owner will use household labor to exploit his own land.

Arrendamiento occurs when a landowner gives land to another in exchange for a sum of money, paid annually. The contract is renewable on an annual basis. This is an infrequent arrangement due to the need for cash in a sum that is difficult to gather at one time. It is often associated with outside ownership; the owner will reside in Lima or in Cuzco for example and give his land in *arrendamiento* to a *comunero*. Land exchange in this way is also subject to other labor arrangements; the *arrendatario,* for example, or individual who pays rent on the land, will often live outside the community and pay wage labor to exploit it.

Migrants to other towns are faced with the problem of the disposition of the plots they leave in the community. Most often, they will simply leave the plot with a family member to cultivate as his own, but more often other formal arrangements are made. *Prenda* occurs when a migrant leaves his land to another individual to cultivate; the individual then sends money or produce periodically to the migrant, based on a common understanding between them.

The most common form of indirect exploitation is variously labeled *a medias, en sociedad,* and *en compañía,* with the latter the most frequent usage. Here the owner will make an arrangement in which he will supply the land and another individual will supply seed, fertilizer, and all the labor. The harvest is divided equally between both partners. This practice is relatively common since it is adaptive to various problems that emerge in land use. The person with an excess of land, for example, will often enter into an *en compañía* agreement with someone who is landless or land poor. While there are some limits to its practice in communities like Rumipata that have large quantities of land in the puna zone, it is a mechanism that reorients land use to an equilibrium between available land and the amount of land that can be cultivated by individual households. As will be seen, the *en compañía* arrangement is also convenient in the sharing of other resources such as capital and information and in the reduction of risks incurred in crop experimentation.

The equalizing effect of indirect arrangements is of course only effective within certain limits. As previously mentioned, in the puna zone of the larger and richer communities, such as Rumipata, some land goes uncultivated because there are simply not enough individuals willing to enter into indirect arrangements. In Tukiwasi, on the other hand, land is so scarce that there is no excess land to warrant indirect exploitation. And in all the communities, pampa land is such a secure and productive resource, capable of being exploited with labor-saving technology, like the tractor and the plow, that indirect arrangements are rare. This latter point is exemplified by a census made in Rumipata to determine the extent of indirect versus direct usufruct arrangements in traditional pampa maize cultivation.[1] The results are presented in Figure 3–2.

Figure 3–2. Usufruct Arrangements in Maize Cultivation, Irrigated Land, Rumipata

	Hectares	Percent
Direct	57.75	83.4
Indirect	11.50	16.6
Total	69.25	100.0

Agricultural Practices

The activities associated with agriculture can be conveniently broken up into the categories of tillage, which may be defined as any activity related to the working of the soil or the crop in connection with agriculture, construction, and animal husbandry. Irrigation, as an activity, will be included under tillage, but the construction and maintenance of irrigation works will be included under the heading construction.

1. This is one area of the peasant communities census where misreporting occurred in response to questions asking the amount of land in indirect cultivation. Peasants consistently underreported land in indirect cultivation because of fears that this kind of usufruct would be grounds for expropriation. This was the reason for the on-the-ground census of pampa plots in Rumipata; it would have been virtually impossible, though, in the puna, given distance, fragmentation of plots, and the peasants' suspicions.

With little exception, the entire farming population of the Pampa de Anta is engaged in one of more phases of the agricultural cycle at one time or another. The only exception is the activity, such as repair and maintenance of homes or seasonal migration, that is carried out during the dead season following the last harvest and ending with the preparation of new land. The outline of the agricultural cycle is given in Figure 3–3.

Major tillage practices include preparation of the soil, seeding, adjustment of the depth of plantings (*lampa*), weeding, and finally, harvest. This basic sequence holds for both the pampa and the puna. As such it is closely tied to the wet/dry seasonal cycle; although the pampa is provided with a permanent source of water, only in certain cases, which will be described in passing, does cultivation extend beyond the maize harvest. The fallow period in the pampa is due to a lack of acquaintance with crops that are frost resistant rather than to a lack of water.

Technology is based on a mixture of an indigenous tool kit, dating to before the Incan expansion, and selected technological innovations that have been successfully adapted to the local environmental situation. The traditional tool kit is made up principally of the *chakitaclla*, *lampa*, *kuti*, and the *ujana*. The *chakitaclla* is a foot plow pushed into the ground by a jumping motion and then pulled back as a lever exposing a piece of sod. This operation is performed by a group, or *masa*, of usually three but sometimes up to five individuals: two men wielding their *chakitacllas* and one man turning the sod over with a pickax to expose the roots. Soil is initially prepared in this way and then left to expose the roots to the rays of the sun. Seeding is performed with a *chakitaclla* or a *ujana* depending on the crop. Root depth must be adjusted for both maize and tubers during the growing cycle; this operation is performed with a *lampa* or *kuti*, a hoe-like tool. Depending on the type of crop, harvest will draw upon the *chakitaclla*, *kuti*, or the *lampa*.

While the subsistence wet/dry cycle and a traditional technology would seem to imply a relatively static peasant economy, there have been major introductions of Western technology, beginning during the colonial period and continuing until the present. Oxen and plows were introduced early

Figure 3–3. Annual Farming Cycle

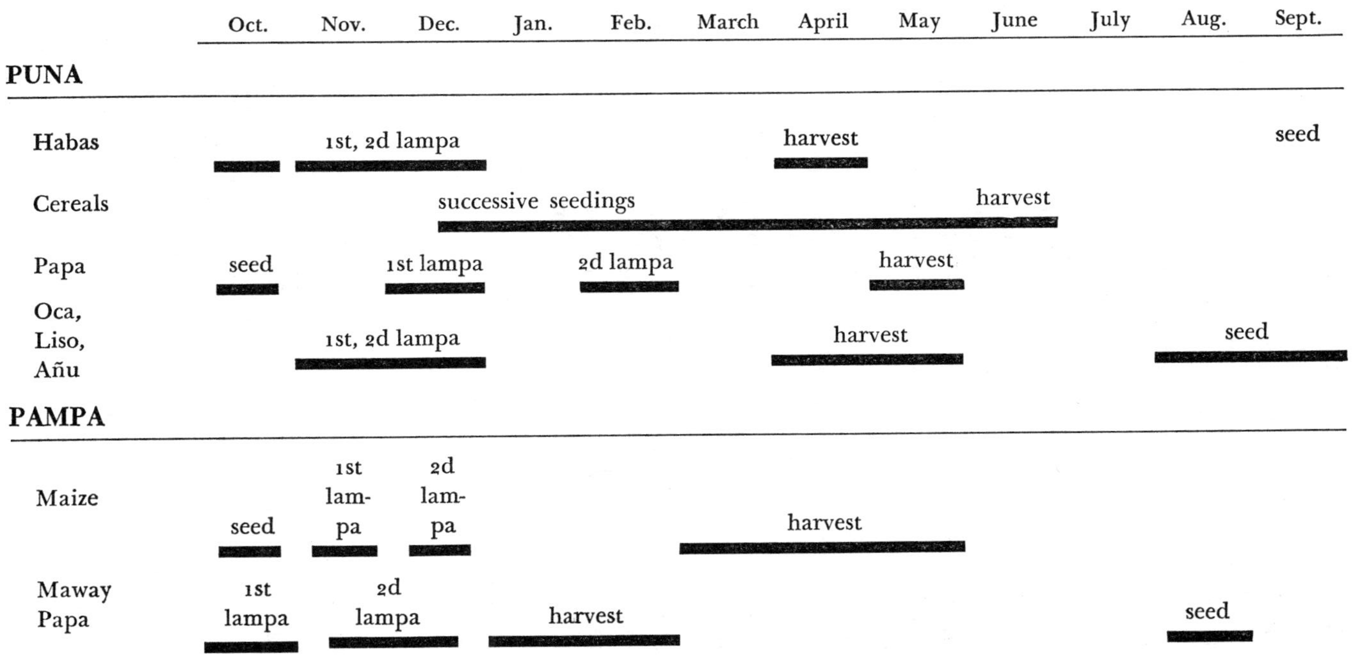

	Oct.	Nov.	Dec.	Jan.	Feb.	March	April	May	June	July	Aug.	Sept.
PUNA												
Habas		1st, 2d lampa					harvest					seed
Cereals			successive seedings						harvest			
Papa	seed		1st lampa		2d lampa			harvest				
Oca, Liso, Añu		1st, 2d lampa					harvest				seed	
PAMPA												
Maize	seed	1st lampa	2d lampa				harvest					
Maway Papa	1st lampa	2d lampa		harvest							seed	

into the pampa agriculture and have drastically reduced the labor and time required for peak periods of soil preparation. During the last two decades, tractors have been introduced, making cultivation even more capital intensive. These technological innovations are a considerable improvement over the *chakitaclla* for soil preparation; they have made maize, for example, less subject to the need for precise water control than with *chakitaclla* technology.

While plow and tractor technology have been introduced into the pampa zones, the puna is still exploited, using the traditional tool kit. The basic reason for this is the fact that the plow and tractor are not able to negotiate the slopes of the mountains, where the *chakitaclla* is perfectly adapted. As a result, in the puna, peak labor requirements during the agricultural cycle must still be met by additional human rather than machine labor.

Technological innovations are one aspect in the dynamic side of the peasant economy; another aspect is the response required by the household head to the variables of climatic uncertainty. This response takes the form of strategies designed to minimize uncertainty; alternate strategies in the case of unexpected actions; and the constant weighing and adjusting of traditional and modern agricultural practices in order to increase production for the household.

Until recently, when export priorities and depleting reserves forced a natural fertilizer, *guano de la isla*, off the regional market, peasants were accustomed to buying it for both pampa and puna agriculture, where it was used in particular during the seeding of tuber crops. It was subsequently replaced by a chemical fertilizer manufactured in Cachimayo, a fertilizer factory located in the pampa. Informants were continually comparing the new chemical fertilizer with the old *guano de la isla* as to their benefits and disadvantages. Chemical fertilizer, they said, did not last as long as natural fertilizer and resulted in an increase in weeds. They said that natural fertilizer from domestic animals was now rising in value and some informants had sold the contents of their corrals to wealthier peasants who would use it as a replacement for the older *guano de la isla*. Even location of homes and corrals is manipulated for optimum soil fertility. Corrals are moved every year or so by some informants and the advantages of relieving oneself in the fields next to the home are well recognized by the peasants as increasing the fertility of the soil. A number of other "organic" practices are followed such as companion cropping, intercropping of *habas* and *arbejas* with maize, and the growing of an aromatic herb around the edges of the pampa plots.

Besides these basic strategies, a number of alternate strategies are employed in cases of unexpected occurrences. For example, in the case of maize cultivation in Antapampa, the usual seeding is from 25 August to 15 September. If a particular crop that is seeded does not germinate due to poor seed or climatic conditions, then a second seeding is done as late as the first week in October. In the second seeding, a special seed, called *Pucutu,* is used; although the yields from this seed are reported to be lower, it is more resistant to frost. If frost continues into fall, as late as 15 October, then wheat and/or barley, which is frost resistant, will be seeded. If a heavy frost continues into fall, then there is no recourse.

The agricultural practices that have been described all represent a set of practices associated with a basic subsistence pattern consisting of potatoes, maize, *habas,* and to a lesser extent dietary supplements such as the indigenous cereals of *quinoa* and *caninua,* and the indigenous tubers oca, *olluca,* and *lisa.* These latter supplements to the basic potato-maize-*habas* complex are not grown to a great extent. Post-conquest crops such as barley and wheat, on the other hand, are grown on occasion in quantity. The basic principle governing crop selection is that sufficient land is seeded with staple crops to meet the needs of the household for the following year.

An alternative to the subsistence pattern that has been presented to this point is the recent trend to the cultivation of onions for market. Although onion cultivation is esablished in only one large size community in the Pampa and is in its initial stages in Rumipata, its implications are particularly important given the context of onion diffusion into the microregion. Further, it presents a case of peasants responding to an agricultural innovation where means have reduced the uncertainty associated with its adoption.

Onions and Urbanization

Onions have traditionally been grown in Cuzco Department of southern Peru in two communities, San Jerónimo and San

Sebastián, located on the outskirts of Cuzco on the south. Due to a favorable ecological setting and permanent irrigation from a river flowing through Cuzco, a classic minifundist economy developed, based on onion cultivation and sale to Cuzco townspeople.

During the 1950s and early 1960s, the situation began to change dramatically. Following a growth of 26 percent between the census years 1940 and 1961, Cuzco began to spread outward and to the south. San Jerónimo and San Sebastián in almost stereotypic fashion became the battleground much like contemporary Miami and Los Angeles for developers intent on bidding for land that could be turned into *urbanizaciones* or middle-class residential communities made up of commuters to Cuzco. It became difficult for minifundists to justify growing onions in the two communities, no matter how profitable they were, when an eager urban developer or real-estate agent was willing to negotiate a high price for one's plot of land.

Another symptom of the "urban demise" was the increased use of the river as a dumping place for effluents from Cuzco's sewer system. Onions from San Jerónimo and San Sebastián began to have a reputation for being irrigated with polluted water. Townspeople were rapidly beginning to understand the effect contamination could have on health. Demand fell for onions from the two communities and rose for onions from other noncontaminated sources.

As the supply of onions from the traditional source began to diminish an interesting series of events occurred. Around 1962–1963 an onion grower from San Jerónimo moved to Pucyurac, a community in the Pampa de Anta, bought land, and began to grow onions. His initial attempts were successful; he found that water was plentiful and not contaminated. The ecological setting in many ways was identical to the southern Cuzco valley and thus eminently well adapted to onions. He began to teach other peasants onion-growing techniques, and by 1970 onions had totally replaced maize as the crop chosen for pampa cultivation. Instead of being restricted to the wet-dry seasonal cycle, onions proved hardy during the colder months and are now grown year around.

Pucyurac replaced San Jerónimo and San Sebastián as one of a set of sources for onion cultivation but even more importantly served as the primary source for the Convención Valley. The railway that connected the Convención passed through Pucyurac and the Pampa de Anta on its way to Cuzco; as in the case of seasonal migration, the railway facilitates the transport of another highland resource, in this case onions, to meet the increased demand in the developing plantation zone.

Onion Cultivation in Rumipata

The success of the transition to onion cultivation in Pucyurac was eagerly watched by peasants in other Pampa de Anta communities. Pucyurac is the first major community one meets in entering the region by road from Cuzco; there is a truck stop there and peasants going to market or visiting Cuzco had ample opportunity to see the maize fields of Pucyurac slowly turn to onion and to ask about the change. What was most obvious to even a casual observer was the high market value of onions in relation to maize. Wholesale prices for one half *topo*[2] of onions in 1971–1972 averaged approximately 6,000 to 7,000 soles as compared with 2,010 soles for average maize yields from the same amount of land. Demand has always been high with wholesale buyers willing to bid for a crop in the field or pay even higher prices on arrival at the market at Quillabamba, capital of the Convención Valley. Unlike potatoes and maize, onions were a lucrative enterprise not subject to the vagaries of seasonal oversupply and slump in demand.

In one of the earliest communities to benefit from Pucyurac's example, a Rumipatiño began to experiment with onions about six to seven years ago. These first attempts, which were followed by those of other peasants, involved a crop sequence of *maway* potatoes followed by onions. *Maway* potatoes are seeded early in the agricultural cycle to yield a crop that is harvested in time for the Carnaval fiesta period before the tuber has matured. One major use is in a ritual meal served during the Carnaval fiesta. Introducing onions into the cycle following the *maway* harvest extended the utilization of land and labor in the production of two pampa crops in place of the traditional wet-season maize crop. By 1972, as is shown in

2. One *topo* is equivalent to 3,333 square meters.

Figure 3–4, onion cultivation has made a considerable dent in that traditional maize pattern. By that year the *Maway*-onion sequence was increasing rapidly and in a few instances giving way to "back-to-back" onion cropping.

Figure 3–4. Irrigated Land, by Cultigen, Rumipata

	Hectares	Percent
Maize	69.25	86.6
Onions*	10.75	13.4
Total	80.0	100.0

* Land cultivated either in onions or in a *maway* potato-onion sequence.

Onions are an unusual cultigen in that considerable labor is required for their cultivation. During soil preparation multiple plowing is required to ensure bulbing, and transplanting from seed sown in a seedbed is necessary when seedlings are about fingersize. Onions must be constantly irrigated as they are extremely susceptible to variations in water supply. Overgrowth of weeds is a hazard and must be controlled. Harvest requirements are less than for maize; onions may be simply pulled from the ground and transferred to a market container. Furthermore, in Rumipata the cultivation of onions extended the growing season beyond that of maize cultivation.

When compared with maize cultivation, capital inputs are much greater for onions. Seed must be purchased and large quantities of fertilizer are required. One of the most difficult capital items to accumulate, however, is the knowledge of the cultural practices associated with onion cultivation. Peasants were aware that onions could be grown successfully in the region by observing their cultivation in Pucyurac; they lacked the experience and the knowledge, however, that their neighbors had accumulated.

A second constraint along with greater inputs of labor and capital is the uncertainty associated with growing onions for market instead of maize for subsistence. There is considerable uncertainty about bringing a crop to harvest, given the scarcity of knowledge of cultural practices and the lack of an experiential base against which to predict local success of onion cultivation. From the point of view of distribution, uncertainty is also found. The complexities of market sale present difficulties for the subsistence peasant. Although surpluses have been sold on occasion, volumes are usually low and transactions occur through middlemen in the community. The decision to enter into the market on a large scale, where sale replaces consumption as the production orientation, is an extremely uncertain action for the subsistence peasant. Quantities, accounting concepts, optimum market locations, and other elements of successful distribution strategies must be considered by peasants contemplating production for sale.

An attempt was made to compare the socioeconomic characteristics of onion growers with the total population of household heads using household census data. The results are shown on Figure 3–5. The average age of onion growers, 49.6 years (N = 22) is slightly older than that of the larger population of household heads, 44.8 years (N = 218). No major differences were found in levels of education and the ability to speak Spanish. Onion growers possessed almost four times the average amount of puna land held by the total population, an indication of the breakdown of the redistributive function of the community in this ecological zone. In the pampa zone no major differences in land access were found. We can conclude that onion growers are a more prestigious and influential grouping and in a somewhat better position to take risks.

Initial adopters are not enormously wealthy in absolute terms, however. Their wealth is land lying in the puna zones; since these lands are relatively low in productivity and require enormous amounts of labor, using traditional technology, this wealth is not readily transformable into liquid capital. There is another qualification on wealth in Rumipata. Because of the channeling of available surplus into the *cargo* system, real wealth has been exchanged for the social wealth and prestige associated with the effects of the *cargo* system. Thus, while onion growers are wealthier in terms of puna land and more prestigious in terms of "social capital" they are still constrained by the relatively high amounts of capital and labor required for onion cultivation.

We discussed earlier the predominance of direct usufruct in maize cultivation in the pampa. Without considering the factors involved in growing onions, we would expect the same usufruct pattern to obtain following such a shift. But

Figure 3–5. Selected Socioeconomic Indices of Onion Growers and the Total Population of Household Heads in Rumipata

	Average Age	Percent Bilingual	Percentage of One or More Years Education	Average Size of Household	Hectares Household Pampa	Hectares Household Puna
Rumipata (N = 218)	44.8	50.0	55.0	4.4	.5	1.1
Onion Growers (N = 22)	49.6	50.0	59.1	4.3	.6	4.2

as Figure 3–6 shows, almost half of the onion plots are cultivated in indirect usufruct.

Figure 3–6. Usufruct Arrangements in Onion Cultivation, Irrigated Land, Rumipata

	Hectares	Percent
Direct	6.25	58.1
Indirect	4.50	41.9
Total	10.75	100.00

The relatively large percentage of indirect usufruct arrangements in the initial shift to onion cultivation can be explained as a response to capital requirements and the constraint of uncertainty.

Onion cultivation requires large amounts of cash together with the necessary knowledge of the cultural practices, both of which were scarce during the early stages of adoption. The most common indirect usufruct arrangement, *compañía,* is particularly propitious for the acquisition of these inputs. Through *compañía* the owner of a plot is able to seek out a partner who can provide the necessary cash. In six out of twelve cases of *compañía* the owner of the land sought out an individual who held an occupation with a regular cash flow, such as a small store proprietor, a blacksmith, or a hat maker. In these cases, it was more profitable for the owner of a prospective onion plot to enter into a *compañía* arrangement in the first stages of the transition to onion cultivation when capital is scarce. Even on a fifty-fifty split, the return from an onion harvest is greater than the returns from a maize harvest.

Technical knowledge of onion cultivation is similarly an input that must be acquired like cash, and indirect usufruct facilitated this transfer as well. Some peasants had experience in the use of specialized techniques such as the application of fertilizer, transplanting, and water control; while others had friends and fictive or real kinsmen in other communities whose knowledge of onion cultivation they could draw upon. In my conversations with onion growers it became clear that sharing of expertise, particularly of techniques that could only be learned through close observation in a common enterprise, was an important aspect of the process and that *compañía* facilitated it through the opportunity for matching talents it presented.

Lastly, indirect usufruct is a logical mechanism for reducing the potential losses involved if one's initial experiments fall awry. Again, *compañía* is particularly attractive in this sense. Where indirect usufruct is found in the Andes among small-scale peasants, sharecropping arrangements, such as the *compañía,* are usually preferred over renting (Orlando Fals Borda 1961:83–84; Emile B. Haney 1969: Table IX–8, p. 245; Havens et al. 1965). Aside from their advantage of not requiring large cash outlays as opposed to renting, the partners share equally in the success or failure of the enterprise. Furthermore, these arrangements have historical depth in the Andes. In James Lockhart's study (1972) of the Spanish residents of Cajamarca during the colonial period, he describes the *compañía* as a "buddy" system in which partners held property in common and interacted together on a regular social basis and in so doing pooled their resources, enjoying greater security in a fluid situation. Thus, the *compañía* falls within that class of sharecropping arrangements, which, economists argue (see S. N. S. Cheung 1969), are preferred for their insurance value.

The indirect usufruct arrangements of Rumipata onion growers are a form of dyadic contract that is a basic building block of peasant social structure in the majority of peasant societies that are not organized into corporate kinship groups, such as lineages and clans. There are two key elements in this arrangement. First, they are based on choice, that is, on a mutual recognition that the interests of each are best served in a bond of this nature. Second, they involve, over time, a generalized exchange of goods and services. The terms of the exchanges may be either informal and implicit or, as in sharecropping, formal and explicit, depending on the degree to which a legal code specifies the terms of the contract and if they are legally and/or ritually validated. But where formally constituted, they are often underlain with an informal pattern of generalized exchange (George Foster 1961).

The elements of choice and exchange enable capital and information to be shared and risk reduced in a situation where one has considerable capital invested. We would further expect that their usefulness in this regard would diminish as the returns to capital following sale of onions enables an owner to build up capital reserves, and knowledge concerning onion cultivation becomes disseminated.

It is important to realize that apart from the formal terms of the contract, informal ties have usually been established through customary means such as gifting, visiting, drinking together, and other forms of social interaction. There are thus costs of both a material and a social nature that enter into the formation of these relationships, which are often not considered.

Construction

A second major activity is the construction and maintenance of large-scale public works projects associated with agriculture. The most important activity under this heading concerns the irrigation system in the pampa that originates in either spring-fed streams in the communities or from streams draining the surrounding slopes. Construction involves the planning and digging of new irrigation ditches to extend the system; where water is scarce, efficient use of the system involves allotments for users, taxes, and some sort of infrastructure to organize these functions. Maintenance takes place during the dry season when ditches must be cleaned of trash and widened to ensure a continual and sufficient flow of water.

All of these functions are necessary for this system to be successful. If, for example, one section is not maintained, it hampers the ability to maintain a continual and sufficient flow of water through the system.

Animal Husbandry

A last set of activities associated with agriculture is animal husbandry, which is defined as including all the activities associated with the care, breeding, and obtaining of produce from animals. The most important activity under this classification is the herding of dairy cattle and/or oxen. They, along with other domestic animals such as pigs, chickens, and guinea pigs, or *qowis,* are raised as a means for savings or for meeting ceremonial needs rather than ordinary consumption in the home.

Most of the practices associated with keeping cattle are

oriented to the acquisition of forage. I have already mentioned the differential resources of the pampa and the puna in this regard; from this it should be clear that the labor involved in the two types of pasture situations is different. In fact, the decision to pasture animals such as cattle and sheep is a relatively complex decision. It involves a number of alternatives that must be weighed by the household head.

The first alternative exists for any *comunero* that has land in the pampa. It is subject, though, to two constraints that have drastically limited its use. First, there is little high-quality pasture around the edges of pampa plots. Usually, forage has to be brought to animals staked out alongside the plot. Second, if an animal is kept in the pampa there is always the possibility it will enter a neighbor's plot and graze on crops growing there. This is the source of many conflicts in the community, and usually ends up in the hands of the *teniente gobernador* who levies a fine on the owner of the offending cattle. There are advantages, though, to keeping one's animals in the pampa: they are usually within eyesight and one does not have to make arrangements that are risk-laden for pasturing animals in the puna. The quantity and quality of milk and milk by-products is much greater from cattle pastured in the pampa.[3]

The second alternative involves arrangements for pasturing animals in adjoining zones in the puna. There are three ways in which they may be done: (1) The owner will move to the puna and build a hut and live there with his animals. This alternative usually is chosen when a person owns or cares for a relatively large number of animals. (2) The owner lends his animals to a friend or *compadre* who lives in the puna to take care of them. He sends rum, bread, and sugar to the caretaker as payment. This arrangement is called *cuidar.* (3) The third alternative, called *compañía,* involves the owner giving his animals to another to take care of for him. They split in half all the produce, that is, milk and cheese, and share the offspring in the following manner: the owner gets the first offspring and the other person the second and so on. The *cuidar* arrangement means, in effect, that the owner must have the cash

3. The advantages are so much greater that from April to September, following the maize harvest, all of the cattle population of the community can be found tethered next to the homes of their owners. During this period there is ample pasture.

to pay for all the store-bought items required in the agreement. This arrangement is preferred, for instance, by a store owner, who keeps cattle as a secondary source of income after the store, which represents his primary income. A peasant with little monetary capital preferred the *compañía* arrangement because it required no cash expenditure, and the other partner had a stake in the welfare of the animals.

Another arrangement that was common during the hacienda period is *yerbaje*. This arrangement is made between a peasant and the owner or representative of an adjoining hacienda. In return for permission to pasture one's animal or animals on hacienda lands, the peasant, or *yerbajero*, who enters into the arrangement agrees to pay a sum of money to the hacienda.

Organization of Production: The Household

The household head in meeting subsistence needs will employ a strategy or a set of strategies that he subjectively assesses as optimum. These strategies involve the allocation of the factors of production (his particular allotment of land and technological resources) at his disposal throughout the period of the agricultural cycle. They are basically short-term production decisions designed to meet the subsistence needs of the household during the following year. All of the set of strategies employed by each household, taken together, can be said to set the basic parameters of the total economy of the community.

In deciding on a set of strategies with respect to the exploitation of land and other resources, a household head will invariably consider the labor of the household as a given; no cost or wage will normally enter into the calculation of labor as an input. While there are inputs that are purchased, such as fertilizer, insecticide, and specialized implements, these are qualitatively different from household labor.

Major tasks such as preparation of land, seeding, weeding, and so on are performed by men. Women and children do not ordinarily play a role in the more important agricultural tasks. For the most part, women are engaged in chores in the home or will assist in herding. If they remain at home they may weave, talk to one another, tend a planting of herbs next to the home, or sit within view of the fields guarding against the ever-present threat of thievery. Their most usual contribution to the agricultural cycle is to bring food and drink to their husbands, especially if there is a work party and the household is responsible for the *hurk'a* or the rum, chicha, cigarettes, and cocoa, consumed during the mid-morning and afternoon breaks. Children do not usually take part in agricultural activities except during periods of peak labor, such as harvest, when all household members, young and old, will participate. Otherwise, children will attend the community school.

While ideally the labor of the household will be sufficient to meet ordinary demands of the agricultural cycle, there are a number of situations in which additional labor must be recruited. Where technology continues to be traditional some tasks will invariably require extraordinary quantities of labor that the household is hard pressed to provide. For example, *barbecho*, or the land-preparation phase of the agricultural cycle is especially crucial since it must be coordinated with the onslaught of the rainy season. Any delay in the *barbecho* can harm prospects for a successful harvest. Thus, in the puna where technology remains traditional, the *barbecho* places demands on the household to provide enough men for a *masa* of three to four persons to wield the *chakitacllas* used in soil preparation. In the pampa, on the other hand, the plow/oxen complex and the tractor have dramatically reduced labor requirements for this phase of the agricultural cycle.

Besides time and technology there are other constraints that enter into labor recruitment patterns. In the pampa, a household usually has no more than one or two plots located next to or less often at a small pace from the house. An individual spends relatively little time engaged in walking to or from his plot. In the puna, on the other hand, an individual may hold up to thirty separate parcels dispersed among the various *tiray* sections. Depending on the sections that are cultivated in any given year, and their location, just to arrive on foot to a section will require from twenty minutes to an hour and a half. The size of these plots furthermore is quite small; informants state they range in size from one fourth (a *checki*) to one half or, in rare cases, as much as a *topo* of land. Therefore, a large proportion of the work involved in performing a task is spent in getting to and from the plot.

A much more efficient utilization of labor, which is the most prevalent pattern in the puna, is the pooling of the labor of a number of men. This occurs as a function of reciprocity between households, which does not stop with the exchange of labor but, as will be seen, extends to other aspects of the local economy. Labor exchange involves the exchange of one's labor under a set of special forms.

In its most simple form a dyadic contract (Foster 1961) will be established by mutual consent between two individuals. It will endure as long as the individuals deem it necessary. Its contractual basis (it is referred to as *wayka*) is informal. In one case that I observed, two individuals worked together on the plot of one in the morning and of the other in the afternoon. Their claim was that their work together was more efficient than one working on the plot alone.

A more detailed form of reciprocal recruitment involves the recruitment of peasants to a work party. For instance, the head of a household, faced with an *aporke*, or soil depth adjustment, in the puna, will visit homes of his neighbors a day before and ask them to join with him in a work party. Those that accept, and not all do, will meet the following morning with their tools and walk together to the work site where they quickly finish the task. Usually when the owner's field is finished the group will then move on to another's field and so on the same day until all of the fields are finished. If the reciprocal obligation is not finished that day, then the party members have the right to expect work from members of the party who have not fulfilled their obligation. Reciprocity in this sense is defined as time paid for time worked; in all cases, the person who recruits the work party is obligated to supply a ration of rum, *chicha*, cigarettes, and cocoa, at around ten in the morning and three in the afternoon, and a meal, the *merienda*, at noon. The recruitment of labor in this form is referred to as *ayni*.

Another type of reciprocal labor occurs in the construction of homes. In this form of recruitment, a prospective home builder will invite individuals to participate in a work party to construct his home. Those who agree will bring their tools and meet on the appointed day at the site. A highly organized operation then ensues. In the making of adobes, for example, three men will mix mud with straw and water, while another will bring the mixture to a man who stuffs it into a mold. This type of work party, although referred to as *ayni*, is different from the agricultural work party. In the house construction party, the spouses of the male members of the work party bring food and prepare a festive meal. The person recruiting the party will supply a *hurka* during the morning and afternoon breaks. In this type of labor there is a festive air to the occasion that is missing from the more utilitarian agricultural *ayni*. It is disappearing in the Pampa de Anta as specialized masons are increasingly being contracted to build homes and other specialized buildings. It appears to be a survival of a festive labor form with indigenous origins, the *minka*, that as in other parts of Latin America, is disappearing with the spread of paid labor (Charles Erasmus 1956).

A third type of labor recruitment involves the contractual recruitment of individuals in which labor is exchanged for cash or kind, known locally as *jornal* labor. This kind of labor recruitment is restricted to those individuals who have access to cash reserves or a substantial agricultural surplus at hand. In practice, *jornal* labor is recruited almost exclusively by the mestizo stratum of the community for basically two reasons: (1) Usually this group has access to cash reserves through the cultivation of cash crops, running a small store, or engaging in wholesaling activities. (2) A positive value is placed on recruiting *jornal* labor by mestizos who find it demeaning to labor alongside of subsistence peasants. Until the rate for *jornal* labor changed as a result of legislation in 1970 the ongoing rates were ten to fifteen soles a day plus a *hurk'a* in the morning and afternoon, and a noon meal. In this type of recruitment, labor is recruited from a "floating" population of the community: the seasonally unemployed, the land hungry, and the landless. Recruitment of *jornal* labor has been the primary source of labor for the needs of mestizo middle holders.

Feudatarios were faced with an additional set of constraints deriving from their relationship to the hacienda. In essence, they participated in two analytically separate economies: that of the hacienda sector in the sense of providing a pool of labor, and that of a subsistence peasant economy, built around the cultivation of plots held in usufruct from the hacienda. The two economic systems were mutually interrelated through the channeling of labor from the peasant household into the hacienda system of production. The labor demands were quite large: in Tukiwasi, five days a week were required by the hacienda, while the remaining two days, usually Saturday and

Sunday, were left for the cultivation of the subsistence plot. Further, the quality of these plots was quite poor; while some were located in the pampa, the majority was held in the puna in a two-year sectorial fallowing system that did not allow land to adequately regenerate itself. In general, the differences between the organization of production at the household level between *feudatarios* and peasants residing in independent communities were differences in degree rather than kind; *feudatarios* were faced with a different mix of production inputs, namely, labor and land, but organized their allocation in much the same way.

We may now summarize the results of our description as it relates to the dynamics of the peasant household. Household heads normally make quite complex decisions in the allocation of their production inputs. Production strategies are short term, seeding land in each of the ecological zones with a mix of staple crops the household head subjectively estimates will be sufficient to last until the next harvest. He does not ordinarily orient production to sale; if his subsistence needs are met, he will then market the surplus through one of various channels available to him. In marketing his surplus he will be quite aware of supply-and-demand factors as they affect prices and adjust his marketing strategies accordingly. He does not ordinarily consider market forces, though, in the planning of his production strategies.

There is a trend, noted in our discussion of the diffusion of onion cultivation, to a mixed pattern of cash cropping and subsistence cultivation. This indicates that peasants will respond to economic incentives coming from the larger society, but only if the payoff is high and there are mechanisms available for reducing risk to manageable proportions. Sharecropping arrangements appear to be a particularly adaptative mechanism for risk and resource sharing. Cash cropping, though, has not yet changed the basic subsistence pattern of the economy of the majority of peasant communities in the Pampa de Anta.

Organization of Production: The Supra-Household in Independent Communities

While the household is the basic unit of the peasant economy, there is another element, the supra-household, which

Figure 3-7. Diagrammatic Representation of the Organization of Production

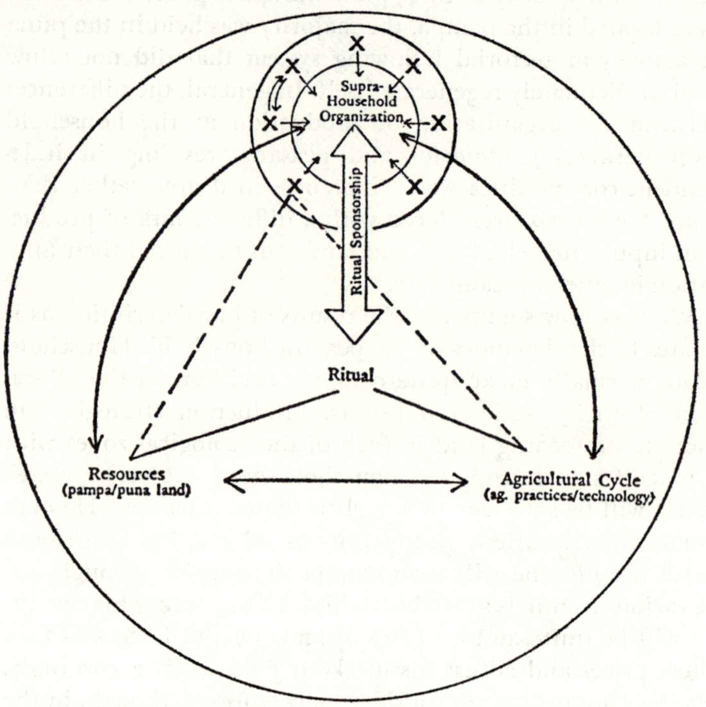

functions in independent communities in a number of ways to maintain the viability of the productive capacity of the household over extended periods of time (see Figure 3-7).

The individual household allocates the resources of land and labor at its disposal in meeting short-term subsistence needs, until the beginning of the next agricultural cycle. Often a household will find itself short of an essential production input, such as land or labor. Indirect usufruct alleviates the problem of variations in land supply, and labor exchange between households is a common means of meeting labor requirements during peak periods of the agricultural cycle. Inter-household exchanges of this sort can be handled on an individual basis and lead to ad hoc alliances of a social nature that last as long as the economic function persists.

The surpra-household sphere refers to the activities of the collectivity of households that are oriented to the long-term as opposed to the short-term interests of the individual household. Formal economic decisions are made in the assembly. The assembly brings together on a periodic basis the heads of the households living in an independent community. Various issues, not all of them economic, are discussed here. Decisions emerge through a tedious process based on normative rules that include face-to-face discussion of issues and agreement through consensus.

What specifically are the functions carried out by the supra-household? They are several in nature and refer to the delicate equilibrium between population, resources, and technology that must be maintained in order for the subsistence needs of individual households to be met. There are two important ways in which the equilibrium can be changed. The first is related to fluctuations in population and the resultant changes in the man/land ratio upon which minimum subsistence levels for the household depend. The second is the necessity to renew constantly resources such as land and technological inputs that will degenerate if left unattended. Such crucial resources include land, irrigation works, roads, and the like.

The Man-Land Balance

One of the most pressing needs of the economy is to maintain the precarious balance between the number of households and the available land to ensure that each household has an adequate supply of land on which to subsist. There are a number of ways this balance is adjusted. It may occur internally, through an annual redistribution of land; externally, through negotiation to acquire new land or to retrieve land alienated from community control; and through the active defense of land from threats of alienation. In independent communities, the *personero* has been traditionally called upon to carry out these duties through the power delegated to him.

Land redistribution occurs in Rumipata and Antapampa through an annual meeting of the assembled household heads presided over by the *personero*. This meeting takes place at the beginning of the clearing of lands located in the puna. The group assembles at the location of the section that is scheduled to be cultivated that year. The *personero* asks if there are any "unowned" lands due to a death leaving no heirs or out-migration. Land reverts back to the community in these instances and the *personero* then assigns it to the households having no land in that sector.

In other instances the supra-household organization will act through its representatives to obtain or maintain land as an adjustment to demographic growth.

In Rumipata, for example, there has been a considerable amount of negotiation between representatives of the community, hacendados, mestizo middle holders, and the courts, over the return of land that has been alienated from communal control. Most recently, successful negotiations between representatives of the community, a middle holder, and a priest living in Anta, resulted in the return of a parcel of land located in the community that had been sold by its owner to the priest for a ridiculously low sum of money. Through persuasion and diplomacy, community representatives were able to purchase the land at a fair price from the priest.

On other occasions communal representatives have represented the community in even more distant and socially distinct contexts. In a famous event of almost mythical proportions, informants recount the struggle in the 1930s and 1940s of the Rumipatinos and their representatives to regain the land that had been alienated by Ezequiel Luna, one of the most powerful hacendados at the time in the Pampa. Only after having made a trip to Lima to speak with the president did a *personero* finally manage to secure a written pledge guaranteeing the inviolability of the communal lands.

A further instance of a community's response to demands on its resource base is in the negotiations made between the representative of the hacienda administrator, and the community of Rumipata in which some irrigated land in the Pampa was leased to distribute as it wished for a sum of five hundred soles a year.

Renewability of Resources

The supra-household sphere of the economy is especially attuned to the maintenance and renewal of its resource base, which, if left unattended would degenerate and thus threaten the security of the individual household. It does this in two ways: by setting rules regarding resource exploitation and by constructing and maintaining large-scale inputs into the technological base of the economy.

Rules regarding the use and exploitation of land are based on each of the communities' evaluation of the ecological parameters of pampa and puna land. The sectorial fallowing system, crucial to maintaining the fertility of puna land, is a prime example. Cultivating out of sequence and with no regard to the rotation cycle can have disastrous effects. Another example is the variation of rules regarding ownership. Pampa land, which does not require a fallow or rotation cycle is allowed to be bought or sold on the open market; while puna land, if alienated from community control and sanction, would be subject to detrimental practices such as over-grazing, reforestation, or disregard for the sectorial fallowing cycle, any one of which would endanger the regenerative and cyclical aspects of puna agriculture.

In general, these rules have evolved over a considerable period of time. In only some instances are they the subject of regular directives. One case in point are decisions made in assemblies to regulate the major steps of the agricultural cycle. For instance, communities must wait until there is a decision of the assembly before beginning the *kachiloc*, or maize harvest. Another case is in the economic activities of the political officers of the community who are expected to conform closely to tradition. In 1971 one of the charges made against an individual who was directing the annual redistribution of land was that he broke the rules of the rotation sequence in the puna by allowing peasants to clear land in a sector that was to have lain fallow that year.

Assemblies provide a forum for the discussion of policies, technological innovations, and other activities originating in the larger society, that, actually or potentially, affect the various parameters of the local ecosystem. In fact, it is difficult for one to maintain an individual stance on these issues. They are debated in day-to-day social intercourse and formally in assemblies. The perception of these issues is strongly influenced by the pressures to consensus.

The production calculus of the household depends directly or indirectly on technological inputs such as roads, irrigation systems, and earth works. Just as important, if not necessarily "economic" in nature, are public works constructed and maintained in benefit of the community. These usually include schools, churches, cemeteries, and the like.

The supra-household is instrumental in recruiting labor for these and other labor intensive public works projects. When a need for considerable amounts of labor is felt, a vote will be taken in an assembly and a *faena* called. A *faena* is a work party in which all of the male household heads are required to contribute their labor for the amount of time specified by the assembly. Female heads of the households do not have to contribute labor in person but are encouraged to send one of the male members of the household or contribute in food or drink.

Failure to comply with the call to work in the *faena* makes one subject to formal or informal sanction. Individuals who are consistently absent from *faenas* are verbally chastised in the assemblies. Assemblies are forums for the public sanction of individuals who consistently break rules governing community life. These most often include failure to attend assemblies and *faenas,* and public drunkenness, and disregard of rules regarding resource exploitation. When these activities occur, public cognizance of them is often sufficient to sanction the individual, but sometimes formal threats of forcing the individual to leave the community or the levying of fines occur. These formal and informal sanctions are an important mechanism for controlling the behavior of peasants that threatens the rules of the community regarding resource exploitation.

The ability of the assemblies to organize and carry out *faenas* that successfully mobilize all of the resident *comuneros* is obviously crucial to the function of the supra-household organization in maintaining the viability of the household unit. In this regard, it is possible to observe examples of reorganization made in response to population growth.

Rumipata decided in 1967 in an assembly to divide the community up into four sectors in order to organize itself more efficiently for *faena* labor. Each sector contained approximately one fourth of the population and was defined in terms of natural divisions of the community. Each sector was named and with one exception (center sector) all of the names referred to long-standing traditional divisions in the community. Each sector takes turns in successive minor *faenas,* and a major *faena* will involve all sectors. There is an elected representative from each sector who reports to the president of the assembly and is responsible for notifying sector members of future *faenas*. All projects undertaken in *faenas* are organized in terms of the four sectors. For example, when a cemetery was planned in an assembly to be constructed on community land, each sector was charged with constructing one of its four walls. When I inspected the site with the president, he carefully pointed out that some sectors were much more compliant with their responsibilities than others.

A larger perspective of the sector organization reveals a close correlation between sector and the concentration of last names, *apellidos*. Certain families live only in a peculiar sector and each sector has a characteristic identity. *Kolgaki* sector, for example, has a number of *misti* families who have become *pequeños propietarios* to a large degree and also has a large percentage of family members in Lima. *Kolgaki* has always participated minimally in the affairs of the community. In retrospect, these sectors are a reorganization of the community along lines that closely resemble the fourfold *suyu* division of the pre-contact *ayllu*; that is, they are residential groupings of related kinsmen that interact on a more intimate and recurrent basis than with members of other sectors or with members of the community at large. This recasting of the *suyu* organization is a direct result of population growth that has weakened the ability of the assembly to maintain and carry out its supra-household responsibilities.

A second instance of community reorganization occurred in Antapampa approximately twelve years ago. Following an order from the offices of the Anta parish to clean and repair all of the churches in its jurisdiction, a conflict emerged between the then *personero* and the president over the location of the church. A small stream divided the community into two halves: an upper and a lower Antapampa. Because the church was located in lower Antapampa the residents of upper Antapampa decided that they too should have a church and began to build one. The dispute arose when the residents of lower Antapampa removed all of the colonial paintings and artifacts and hid them in the home of the president and the school. The upper Antapampeños asked that they be returned to the church and the request was denied. The conflict ended up on the desk of the archbishop of Cuzco who ruled that the paintings and other artifacts would be divided up equally between both groups. Since that time, there has been a reference to Antapampa Chico and Antapampa Grande divided by a river between the two communities. Besides this geographic division there are social divisions, Antapampeños refer now to Antapampa Chico as *hawakollaykuna* or 'tall cactus' because of the persistence of cactus in that sector, while Antapampa Grande residents are referred to as *alkowanukuna* or 'dead dogs,' because its location at the side of the highway makes it the scene of many dead dogs killed by trucks.

Organization of Production: The Supra-Household in Haciendas

The major difference in the organization of peasant production at the supra-household level between independent communities and haciendas is that *feudatarios* lack the autonomy to make and enforce collective decisions regarding the long-term state of their factors of production, namely, land and labor. Instead, all decisions of this nature are made by the hacienda administrator. The assembly in hacienda communities is called by the administrator to inform *feudatarios* of their duties. He assigns subsistence plots, sets, and enforces rules regarding the use of land, and, if he is concerned about the welfare of *feudatarios* may negotiate for land from adjoining communities or haciendas, as did MCS in obtaining land for Tukiwasi from Rumipata. He acts with the power delegated to him from the owner of the hacienda. The ultimate manifestation of that power is his ability to have a *feudatario* expelled from the hacienda and thus from his source of subsistence.

Summary

We have isolated in this chapter two spheres of the peasant economy: the household and the supra-household. In the first sphere, the head of the household ordinarily makes quite complex decisions as to the allocation of his stock of production factors in meeting subsistence needs until the next agricultural cycle. He is oriented to security, but as our discussion of onion diffusion reveals he will take risks that have a high return, if sufficient mechanisms exist for reducing uncertainty.

The supra-household sphere is the long-term correlate of

the household. In the independent communities it functions through the assembly of household heads and those individuals to whom it delegates power in its behalf. Long-term policy in the assembly is based on the collective decision of those present. In the hacienda community the *mayordomo* makes all long-term decisions affecting households based on power allocated him by the hacendado as well as other power stemming from his position as a mestizo in the regional political economy. In the independent communities, reciprocity, redistribution, and renewability of resources are normative principles that underlie the economic behavior of the supra-household in maintaining the parameters of the local ecosystem; the latter include the delicate balance between population, resources, and technology essential to the survival of the household.

4

Origins and Processes of the 1969 Agrarian Reform

The Agrarian Reform decree-law that was passed in 1969 attempted to rectify the defects in the agrarian structure of the southern sierra, such as the ones described for peasant communities in the Pampa de Anta, as well as for other parts of Peru. It was a national reform policy that arose out of a series of events entirely outside the perspective of the Pampa and of the limited set of interacting groups located there. To understand its impact upon the political economy of the Pampa, and, in part, the eventual response of the peasant beneficiaries to it, my perspective will now shift to the national level, and I will also discuss the series of events associated with the genesis and eventual implementation of the 1969 Agrarian Reform.

One of the basic aspects of the uncertainty problem vis-à-vis peasants is that often new economic alternatives originate in the national society within which the peasant sector is a dependent appendage. Because of this, it is important to understand the motivation behind the extension of new economic alternatives to the peasantry; if part of the motivation stems from pressure, either direct or indirect, from peasants, then one element of uncertainty, that is, the unexpectedness of the alternative, is reduced. Similarly if provision of the alternative occurs rather independently of the peasantry, then its novelty is increased by the element of surprise.

Any subsequent analysis of changes in Peruvian society must first consider their relation to the reforms carried out by the military junta of 1968. To understand the behavior of the military junta, which is difficult to understand given "traditional" models of the behavior of Latin American military groups, it is essential to view its place in the context of changes that have occurred in Peru in the twentieth century, in particular, during the 1950s and 1960s. The military revolution is a response to these changes and an attempt, in its program of the reforms, to reorient the course of change in a direction that is consistent with "national" goals and aspirations as interpreted by the military.

Although the analysis of these changes is extremely complex and open to varying interpretations, it is possible to begin by looking at the major sources of cleavages in Peruvian society that have emerged in this century. They are related to a number of causes including: the shifting distributions of resources upon which the national political economy is based, the interrelated processes of migration and urbanization, the emerging of interest groups with a class base, and a changing ideology stressing nationalism and the equitable distribution of national resources. These factors in one sense may be seen as independent, but a closer analysis sees them as mutually interrelated, as for example the relation between rural-urban migration, the growth of Lima, and the emergence of an urban proletariat. Our purpose here will be to outline individually these factors keeping in mind that the Revolution of 1968 was a response to the cumulative effect of them.

First of all, the revolution was a reaction to the inability of the governing coalition that had reigned in Peru since the beginning of the century to represent adequately the interests of all of the people. This coalition made up of the government and a coastal-based elite had effectively maintained control over the political system of the country. This coalition dated from the middle part of the nineteenth century when profits from the sale of *guano* on the world market enabled some of the landholders in the fertile coastal valleys to invest in capital improvements to their holdings and develop a highly rationalized plantation agriculture. Later, during the first decades of this century an export boom, based on cotton and sugar from the coast, wool from southern Peru, and rubber and oil from the tropical lowlands, further reinforced the economic position of the coastal-based oligarchy. Although there have been differing interpretations[1] as to the source of

1. Two useful sources for the economic history of this period are Macera (1975) and Yepes del Castillo (1972). See Greenhill and Miller (1973) for an account of the shift from guano to nitrate exploitation in the 1870s. The guano era is analyzed by Bonilla (1967–1968, 1974) and Flores Marin (1977) is a good source for the role of rubber exports. The varying interpretations of the nature of the power of the contemporary oligarchy are brought together in a re-

power wielded by the oligarchy there is general consensus that in one form or another it does exist and that it exerts a disproportionate amount of influence in the political economy of the nation.

The inability of the traditional governing coalition of the agricultural and mining interests and the government to represent the new class interests of the middle half of this century is related, as in the rise of the oligarchy in the nineteenth century, to a series of adaptations of the Peruvian economy to events stemming from foreign interests. Peru has experienced a transition in the years since the thirties from an economy based on agriculture and mining, essentially primary sector pursuits, to an economy based on the development of an incipient urban industrial economy based on manufacturing. Agriculture lost its traditionally strong position in the transition, particularly in the sierra where agriculture continues to lose its, at best, marginal role in the national agricultural sector. Along with the shift to an incipient import-substitution industrial economy there has been a concomitant shift in rural-urban relations. According to Anibal Quijano Obregon (1968):

> . . .the figures on present developments in the economic structure show that economic growth is concentrated in the cities, giving them complete economic domination of the rural areas. At the same time, moreover, the deterioration of agricultural production makes these rural areas completely dependent upon the urban areas. Thus, the present demographic relations between town and country by no means imply that the traditional economic rela-

---

cent collection of essays (Francois Bourricaud et al, 1971). For Bourricaud, the oligarchy consists of a nucleus of coastal families that through their manipulation of foreign capital returns from agriculture is able to control effectively key institutions such as credit, import and export mechanisms, and internal investment. Jorge Bravo Bresani differs in seeing the oligarchy as essentially dependent on international corporations for its economic power. In his view, the oligarchy merely acts as the mediator in negotiating the "conditions" imposed by the supra-national interests on the Peruvian government. Henry Favre returns to the internal source of power of the Peruvian oligarchy and sees it as extending its influence through diversification of its economic activities. For Favre, a crucial concomitant of the oligarchy is in its creation of a national capitalist class with narrow class interests at conflict with the nationalistic interests of other groups in the Peruvian society.

> tions between them, which granted the rural areas relative social, economic, and cultural autonomy, persist. In fact, the rural areas used to dominate the towns and cities, since the center of the country's economy was in agriculture and cattle raising [p. 310].

The weakening role of the rural areas is further hastened by the mass exodus to the coastal cities, especially Lima, of migrants from the rural areas. The process appears to have been initiated by the availability of work on the coastal plantations as they expanded from the results of capital investment in the latter part of the nineteenth century. Both the direction and rate of the migratory stream increased dramatically during the last twenty years to the cities on the coast, and to a large extent account for the large increases in population of Lima and the more prosperous coastal cities such as Arequipa, Trujillo, and others. For example, between 1940–1957 Lima and Callao tripled its growth. The 1961 census established that 47 percent of the Lima-Callao population was born outside of these provinces (J. Oscar Alers and Richard P. Appelbaum, p. 3).

To a large extent the transition to an urban industrial economy and the rapid development of the secondary and in particular the tertiary sectors is related to the process of in-migration and concomitant urbanization. Figure 4–1 summarizes the economic shifts during the 1940–1961 census years.

Figure 4–1. Percentage of Economically Active Population by Sector, 1940 and 1961

			1940		1961	Net Change
Agriculture	Primary	62.4	64.4	49.7	51.8	−12.6
Extractive		1.8		2.2		
Manufacturing	Secondary	15.4	17.2	13.1	16.6	− 0.6
Construction		1.9		3.1		
Commerce	Tertiary	4.6	16.8	9.1	27.2	+10.4
Transportation and Communication		2.1		3.0		
Services		10.2		15.2		
Other		1.6		4.4		+ 2.3
		100.0		100.0		

Source: Derived from David Chaplin (1967:169).

As analyzed by Quijano Obregon (1968) the transition in the national economy fostered primarily by the effect of large-scale investment in industrial activities by foreign interests and rapid urbanization are responsible for a number of important changes in Peruvian urban society that have occurred to a much greater and more rapid extent than in the rural society:

> The most important elements (of urban change) that stand out are the following: (1) the consolidation of the tertiary and industrial sectors as the main core of the national capitalist class, implying the rapid elimination from this class of the still remaining social and psychological characteristics of the traditional Spanish and French aristocracy; (2) the numerical increase of the urban "new middle class" as the result of increased bureaucratization and professionalization, and a growing petite bourgeoisie, based upon the expanding commercial sector and small industries engaged in tertiary activities; (3) the increase, especially in the large cities, and most of all in the metropolitan area of Lima-Callao, in the number of factory and non-factory workers; (4) and the growth of a huge mass of unemployed and underemployed people who have no definite place in the newly developing socioeconomic structure except as oppressed outsiders [p. 308].

The strength of the new middle sectors in the political situation of the country became apparent for the first time in the 1962 elections that pitted Haya de la Torre, representing a coalition of the traditional upper class and Alianza Popular Revolucionaria (APRA), against General Odría the candidate of the oligarchy and Fernando Belaúnde Terry the choice of the military. Belaúnde had lost the presidential election in 1956 against Manuel Prado who had used an alliance with opportunistic APRA members to effect a coalition of the upper class and the urban lower and middle sectors. During the interim between 1956 and 1962, Belaúnde founded a new party, the *Partido Acción Popular,* which soon attracted a wide following from the new middle sectors of Peruvian society. His program emphasized reform and contained elements that appealed strongly to the emerging nationalistic and self-conscious middle sectors. Arnold Payne has summarized Belaúnde's platform as follows:

> In general, Belaúnde's campaign programs were the ones he had meticulously outlined in *La Conquista del Peru por los peruanos*: a preference for a mestizaje, or mixed economy; a comprehensive

land reform, a reform of the nation's inequitable tax structure; easier credit; greater autonomy for the provinces and municipalities; a program of military-civilian cooperation and a greater role for the armed forces in the overall development of the country's accelerated program of housing construction; the creation of a new government agency to promote community development which would be called "Cooperación Popular"; and the construction of an international highway along the eastern slope of the Andes (the Marginal Highway of the Jungle) which would not only integrate the entire country through a network of feeder roads but would also open up to exploitation and cultivation thousands of hectares of fertile virgin land. (Arnold Payne 1968:40)

Belaúnde's emergence as president out of the old combination of candidates that appeared during the 1962 elections was taken as a step in the direction of a broader-based government than those that had preceded him, as in the case of the Odría regime that had catered mostly to traditional oligarchial interests. Belaúnde, a young architect with a taste for reform, was carried by the electoral participation of the middle and lower sectors of the electorate and by the support of the military, which was aligned against the coalition of Haya de la Torre and the upper class. His government was essentially middle class and offered clear alternatives to the stagnant situation that had characterized earlier presidencies.

The inability to carry out reforms that were on the minds of most Peruvians combined with a series of reverses in the political arena eventually led to Belaúnde's overthrow in October 1968. During the early years of the regime, the radical left began to organize along the lines of guerrilla bands in order to capitalize on the peasant movements that had stricken the rural countryside during the late 1950s and early 1960s. Belaúnde's reaction at first was to minimize their importance, probably due to an anguished outcry of the upper classes; the peasant movements had increased dramatically almost as soon as Belaúnde had entered office and were interpreted by spokesmen of the agricultural interests as a premediated plan to subvert the social order and as a direct attack upon the agricultural productivity of the country (Sociedad Nacional Agraria 1963–1964:64–65). But after the appearance of organized guerrilla groups in different parts of the country the military began to criticize openly Belaúnde for not letting it put down the movement. The military finally got its way. In 1966 the military completely eliminated the threat from guerrillas by a massive manhunt made successful by the use of counter insurgency training and new equipment supplied by the United States. The military emerged from this experience with a taste of success and as a self-appointed champion of reform. At this point the initial cleavage between Belaúnde and the military began to appear.

Belaúnde's position continued to weaken dramatically in the middle of 1966 because of economic problems stemming from economic stagnation and deficit spending. Decreasing food production and a small volume of exports, particularly cotton and sugar, produced a sizeable negative balance of trade that compelled the government to borrow from foreign sources (Michael Locker 1969*a*:6). In the latter half of 1967, the sol was devalued drastically, by almost 40 percent, after it had been stable for a decade. Belaúnde's efforts to meet the crisis by refinancing the national debt with a group of foreign banking firms primarily from the United States was highly criticized (Karen Spaulding 1969):

> The terms of the loan appeared to many in Peru to be extremely high, and early complaints were turned into charges of complicity and fraud when it was revealed that Deltee Peruana, the Peruvian subsidiary of a U.S. based international investment firm, of which the Minister of Finance was president, was one of the firms contributing to the loan.

The most damaging element in the worsening situation occurred in a series of moves by Congress involving the International Petroleum Corporation (IPC). Belaúnde had promised upon taking office that he would negotiate a new contract with IPC within ninety days, but, after a lapse of five years, an increasingly hostile Congress finally passed a law in July of 1967 that confiscated La Brea y Pariñas and directed the president to negotiate a new contract more in the interests of the country. The Peruvian tax court found the corporation guilty of failure to pay $144 million in duties on past imports of gasoline and the stage was set for the confrontation between United States interests and Peruvian over the manner in which the problem would be resolved. Although Belaúnde subsequently announced an agreement had been reached with IPC, it was soon found to have involved considerable concessions. This fact as well as some shady dealings involving the content of the contract was enough for the inevitable: the armed forces led by General Velasco assumed power in a bloodless coup.

What is perhaps surprising, given traditional interpretations of the role of the military in Latin American countries, is the subsequent behavior of the military junta. Almost immediately upon assuming office the IPC was nationalized outright; later the passage of the agrarian reform law and the immediate expropriation of the major coastal plantations were followed by a new water code prohibiting private ownership of water resources. These institutional changes along with a series of others that are continuing to this day have, as one writer for a leftist review of Latin American affairs put it, "triggered the most radical institutional changes that have occurred in any Latin American country since the Cuban Revolution" (Locker 1969:1:1). Within two weeks after Peru's new land reform law was decreed, Fidel Castro gave his endorsement: "Our judgment of this law is that, if executed fully, can be described as revolutionary. . .if in Peru a true revolution develops, it does not matter that those who promoted it are a group of military leaders."[2]

Two possible factors have been given to explain the behavior of the military junta. First, there is considerable evidence that the makeup of the military corresponds to the emergence of a middle sector in Peruvian society that identifies with middle- and lower-class interests in opposition to the upper class. Luigi R. Einaudi traces the origins of the military in the middle class to the decline in prestige of the military as an upper-class pursuit in the period following independence. Since the turn of this century a military career beginning in the Military Academy has offered a free higher education leading to a respectable profession to members of the impoverished but respectable families in Lima or the provincial capitals. In recent decades this shift has been even more in the direction of the lower classes: "The shift within the officer corps in recent decades has been away from the whitish upper middle classes and towards the darker lower classes. It has been away from the coastal urban centers and particularly Lima, and toward the rural provincial towns of

2. From *The Economist para América Latina* (London), 23 July 1960, p. 10 (quoted in Gall, p. 303).

the interior" (Einaudi 1969:5). Einaudi further sees two important political consequences that follow:

> The first is that most officers experience a sense of exclusion or frustration in their relations with the top elite. They do not come from it and cannot expect to rise into it during even the most successful military career. They tend in fact to be hostile to what they themselves often call the plutocracy. The second point is that the officer corps is sufficiently distinct from the masses to feel no particular identification with them either, with the possible exception of a paternalistic regard for the Indian. (Einaudi 1969:5)

Given the separation, if not isolation, of the military in terms of its social origins from the traditional sectors of Peruvian society, the question still remains of its ideological commitment to reform. To a large extent the answer comes from the self-image that the military has of itself, based on a number of factors: a highly trained professional military presence; its perception of a national role as the spokesman for lower- and middle-class interests; its view of a role as an efficient functioning bureaucracy; and, lastly, the combination of these facets of the military "group personality" into a set of national goals with respect to direction and reform. A major factor in the self-image described here and of the content of the reform ideology espoused by the military has been the role of its Centro de Altos Estudios Militares (CAEM). This institution, created by an act of the Odría government, has been responsible since the latter half of the 1950s for the training of selected younger career officers destined to hold the higher positions in the military hierarchy. Victor Villanueva's important study (1972) of the CAEM reveals that the initial concerns of the institution, to maintain a national defense posture and the general well-being of the country, were broadened following a conference in 1962 to include a distinction between political and national objectives. Further, concepts such as social justice and socioeconomic transformation became means for the determination of national objectives (pp. 90–91). These concepts and a basic strategy were further developed in the evolution of CAEM, and an important factor has been the invitation of civilian academics and professionals to the institution to give seminars and public addresses to the participating military. It followed that political participation was necessary in order to be consistent with its perceived self-conception; this nationalization was added to the reformist ideology of CAEM. The success of the military in the guerrilla campaign of the early 1960s increased its self-confidence and expands its perception of the military role.

Seen in the context of the military's ascent to power and its inclusion of interest groups aligned against the oligarchy, the agrarian reform law that was passed by the military after nine months in office served a dual purpose: first, it was a needed reform that attempted to integrate sectors of Peruvian society by bringing them into the market economy and in the process increasing the productivity of the agricultural sector. Second, it attacked initially the very base upon which the coastal oligarchy rested (and importantly the base of APRA's strength in the north) through expropriation of the coastal haciendas within twenty-four hours after the law was enacted. In terms of the class interests represented by the military and earlier by the Belaúnde interests, the agrarian reform was well received; it was an attempt to focus on the problems in the rural society seen by many as push factors in the mass exodus of peasants to the coastal cities. In addition, there was little sentiment felt for the hacendado class in the sierra; it was a symbol of Peru's feudal past that continued to plague it.

There have been attempts at constructing agrarian reform legislation for the country since 1949, but the first real Agrarian Reform Law (No. 15.037) was enacted during Belaúnde's presidency. This law, proposed as part of his electoral platform, is in retrospect the government's response to the peasant unrest that swept the country during the early 1960s. The ability of agricultural interest groups, especially the strategically placed members of the Sociedad Nacional Agraria, to lessen the impact of the law was considerable. Two major defects plagued it: first, clauses written into the law specifically exempted cultivable land owned by corporations from expropriation. As Rubens Medina (1970:8) has pointed out, the legal status of "corporation" was often a mechanism to cover up concentration of large amounts of land owned by a few prominent families. Eighty percent of agricultural land (fifteen million hectares) was owned by these families. If by some means land was still deemed subject to expropriation, a second major defect, an immense and immobile bureaucracy, considerably reduced the prospects of expropriation. Medina (1970:13) in commenting on the long periods of time involved said: "A total time of one year and ten months had been estimated as the usual period to finish a single process, but the first case took three years and two months and some cases have even been longer."

These two defects, at least during the first two years or so of operation, did not notably mar the junta's agrarian reform. Besides acting with lightning like speed, the military was quite impartial in considering land for expropriation. Both the semi-feudal and highly unproductive haciendas of the sierra and the modern agribusinesses of the coast were judged fair game by the military junta. When difficulties and legal loopholes cropped up, there was a quick response in attempting to solve the problem. There were few such problems; perhaps the most famous was the clause in Article 9 of the original law that allowed one to escape expropriation by dividing up and parcelling out one's land by private initiative. This did lead to some excesses as owners retained control over their properties by including friends and close relatives among the recipients of parcels. The military soon reacted to the loophole, though, by a presidential decree (No. 18003) that effectively eliminated the practice. In a thorough documentation of legal responses to such loopholes, Luis Pasara (1970) has shown the willingness and speed with which the junta overcame obstacles in the implementation of the law.

The major focus of the agrarian reform during the first year were the departments in which there had been prior agrarian problems and the pattern of tenure was most inequitable. These included the coastal Departments of Lambayeque, La Libertad, Piura, and some of the districts of Lima Department. Cuzco, Pasco-Junín, and Puno Departments, where peasant mobilization was acute during the early 1960s, were the first departments in the sierra; later Ancash, Ayacucho, and Huancavelica Departments were added to the list. As Figure 4–2 indicates, well over a million hectares were expropriated during the first year.

There are two basic processes in the implementation of the agrarian reform law. The first process is glossed *afectación* in the legal code and refers to land expropriation. It begins with a declaration by supreme decree of an agrarian reform zone, in most cases a department. Following publication of a formal notice, all owners of cultivable property must present themselves in the office of the local Ministry of Agriculture

Figure 4–2. Land Expropriated, in Hectares, between June 1969 and August 1970, by Departments

	Hectares	Percentage of Total
Piura	4,794	0.3
Lambayeque	78,753	5.3
La Libertad	266,061	17.8
Ancash	64,603	4.4
Lima	23,131	1.5
Pasco-Junín	449,896	30.1
Cuzco	79,194	5.3
Puno	406,861	27.2
Huancavelica	23,574	1.6
Otros	96,691	6.5
Total	1,493,558	100.0

Source: Derived from Pasara, p. 47.

where they pick up a form to fill out that is a formal description of their property. When all the forms are filled out and turned in to the regional office of the agrarian reform, a program is drawn up of the order in which each province is chosen for expropriation. According to this schedule, each province is then analyzed to determine which properties are to be expropriated along lines set out by the agrarian reform law.

The formal media sources are the base for the initial dissemination of the news surrounding the law. Under the provisions of Article 50 an announcement of the decreeing of a zone as an agrarian reform zone has to occur three consecutive times in the local newspapers (*El Sol* and *El Comercio* in Cuzco) as well as posters placed in the public offices of the district and provincial capitals and "by any other means deemed advisable." As will be seen in later chapters, the Ministry of Agriculture went to considerable lengths to use highly sophisticated media sources, such as the extensive use of mobile public address systems and radio programs in Quechua, to reach the beneficiaries of agrarian reform zones.

After a study of the forms filled out by the property owners and an analysis of the socioeconomic characteristics of the zone, the agrarian reform branch of the Ministry of Agriculture issues an official resolution that certain properties are to be expropriated. An appeal to this decision may then be made through a land judge (*Juez de tierras*) and an agrarian tribunal set up to deal with judicial matters concerning land. Eventually a supreme decree coming out of Lima seals the fate of the land in question: signed by the president, it states that a piece of land is to be expropriated and no appeal will be entertained. Following the Supreme Decree a calculation is then made of the net worth of land and other goods that form the bases for compensation to the landowner as required by law.

The second process in the implementation of the agrarian reform is distribution of the expropriated land to beneficiaries. This process is referred to as *adjudicación*. It is based on the recommendation of a committee created by Supreme Decree as to the manner in which the land is to be distributed and is then reviewed by the agrarian reform section of the Ministry of Agriculture. The recommendation in turn is based on a study of the agrarian reform zone including the type of beneficiaries, the natural resources of the zone, and the socioeconomic infrastructure. The beneficiaries are determined by the agrarian reform section of the Ministry of Agriculture. A socioeconomic census is made in the region and provides quantitative data on individual households.[3]

The decision as to the manner of distribution is a result of the recommendation of the study committee and the final review by the agrarian reform section. It may refer to three basic modes of distribution provided in the Agrarian Reform Law by Articles 66 through 90. Under the provisions of these articles, expropriated land may be distributed to peasant communities, cooperatives, and social interest societies (SAIS). Each of these modes of distribution is designed to be optimum for the range of socioeconomic situations encountered in the country. The distribution of expropriated land to individuals is rare and only occurs under extraordinary circumstances. In practice, under the provisions of the law, land is not given directly to an individual to cultivate but to a production unit

3. These censuses, which number about thirty volumes and are published by the Ministry of Agriculture, are a valuable source of data on the pre-agrarian reform situation especially since in many of the regions of the country they represent the first systematic agricultural census that includes data on topics such as landholdings, cattle, and so on, at the family and community level. Until 1972, the previous agricultural censuses were based on estimates.

that controls rights over land use. The production unit refers to the peasant community, SAIS, and cooperative.

Figure 4–3. Composition of Agricultural Units Created as Beneficiaries of the Agrarian Reform to 30 June 1971

Agricultural Unit	Percent of Expropriated Land
Collective	
Cooperative	40.7
SAIS	34.8
Communities	17.8
Individual	6.7
	100.0

Source: Derived from Table 3 in Wayne R. Ringlien, 1972.

As Figure 4–3 indicates, approximately 75 percent of the expropriated land was distributed to either a cooperative or its close variant, the SAIS. This preference for the cooperative mode of distribution is supported by statements made by agrarian reform planners (Garcia, p. 7) and government policies of extending agricultural credit to cooperative production units in preference over individuals. While part of this preference stems from economic ends it is also clear that there is a political element. As has happened in other agrarian reforms in Latin America, particularly the Bolivian case, the government in power had found a means of institutionalizing control over a newly freed peasantry and the rural proletariat by placing beneficiaries within corporate bodies that at some point respond to centralized control at the regional and national levels. The effect has been to divorce peasant and rural proletariat support from cliental policies of the populist or radical variety and to place it under the aegis of state administration and control. This had been the effect, although modified in subsequent moves by the military junta, of meshing cooperative exploitation of the coastal plantations and government administration. Not surprisingly, the coastal proletariat has been a potent source of conflict with the military dating back to the Trujillo uprising of the 1930s that was put down by military repression.

On the coast, where large highly mechanized plantations were the rule, the mode of land distribution has been the producer's cooperative in which all of the land and installations of the plantation are turned over to a cooperative that is essentially made up of the same personnel as before the reform. Depending on the complexity of the enterprise, there may be merely a varying sized group of workers and an intermediate group of administrators. The actual organization of the operation changes little under this mode except for the removal of the owners and their immediate representatives in the administration, who are replaced by administrators hired or selected by the cooperative.

The SAIS is a peculiar kind of cooperative unit that has been applied in the cattle regions of the central and south sierra, notably the cattle regions of Junín and Pasco and the Department of Puno. In effect, the SAIS consists of communities bordering on the edges of the large cattle haciendas that are selected by a process of analyzing relevant census and socioeconomic data.

The direction taken by the Peruvian planners is similar to that advocated by a number of Indian scholars, who have given much thought to the problems of land reform and economic development. Large collective farms, they argue, would make more efficient use of agricultural manpower and would permit the transfer of the surplus labor into public works projects, a form of capital accumulation in the rural areas (V. M. Dandekar 1962). Although the Peruvian planners do not directly give the origin of their emphasis on cooperative tenure, they were certainly aware of arguments similar to those of the Indians, and, further, the solution offered by cooperatives to the introduction of improved technology and technical assistance.

The cooperative model has been given preference not only in the adjudication of properties expropriated in the agrarian reform, but also in other major areas, both rural and urban, of governmental reform. A new Peasant Communities Law (No. 37–70–A), issued by decree in 1970 as an adjunct to the Agrarian Reform Law, substantially modified the organization of recognized peasant communities. Prior to the Peasant Communities Law, "indigenous communities" had a peculiar organization, as outlined in Chapter 3, dating from the provisions of the Constitution of 1920. Under the provisions of the new law, peasant communities now have an organization almost identical to that specified in the General Law of Cooperatives (No. 15260). There is an administrative council, charged with the administration of community affairs; a vigilance council, which overlooks the activities of the administration council; and a general assembly of *comuneros*, the maximum decisionmaking body of the community, which sets long-term policy and reviews the actions of the administrative and vigilance councils. To hold office in either of the elective councils, an aspirant must be literate in Spanish; to hold the office of the president of the council of administration, one must be a voting Peruvian citizen.

The Industrial Reform Law, decreed by the military government, establishes guidelines for the promotion of industry, the treatment of foreign capital, profit sharing, and worker's participation in management. One of the key provisions of the law is a section dealing with reform of the enterprise where mechanisms are set out for an industrial community that represents the worker before the enterprise. The industrial community is based on a cooperative framework and allows the worker to participate in management and profit sharing:

> The Law prescribes that each enterprise will have to set apart, after having paid taxes, 15% of the annual net profits, in order to acquire shares on behalf of the Industrial Community. When the Industrial Community will have acquired 50% of the shares of the enterprise, its members will turn into individual owners of the shares and the profits derived from them.

According to Joost Kuitenbrouwer (1973:19), the "workers in Public Industries will receive bonds instead of shares."

Workers become members of an industrial cooperative that elects representatives to the board of directors of the enterprise. The boards of private and cooperative enterprises have at least one worker's representative. As shares are progressively acquired by workers, the number of representatives increases.

> According to the President, the new General Law on industries . . .provides the workers with an important participation in the utilities and the direction of the enterprise. It alters the traditional system of ownership, giving the workers progressive access to the ownership of the industry. . . .The new structure was particularly meant, as was announced, to "forge a new personality which the workers will gradually acquire when they are no more

> simple workers but creative persons in a human community with which they can really identify as their community." (Kuitenbrouwer 1973:20)

The cooperative model, in effect, is the arrangement that the government has found optimum in meeting a number of goals; it appears to be an efficient economic unit to many and thus allows the delicate process of distribution of expropriated lands to be accomplished with a minimum of loss of productivity. It is amenable to economies of scale and of services by the incorporation of cooperative units into regional and national federations. Profit sharing in industry and in agriculture can be accommodated within a cooperative format. And, importantly, the cooperative allows in principle a maximum of participation of its core membership in the administration and direction taken by the unit. Given the importance participation has in the thrust of post-1968 military planning, the cooperative is an ideal vehicle for attaining this as well as other pragmatic ends.

To summarize, the agrarian reform that was eventually implemented in 1969 was the culmination of a series of socioeconomic changes in Peruvian society beginning in the latter part of the nineteenth century and continuing into the present. The military junta that came into power in 1968 was a reflection of these changes and its subsequent behavior clearly placed the military junta in a peculiar historical role that analysts of Latin American politics are still coming to terms with. The institutional reforms, by decree, have characterized the junta as just as "revolutionary" as any other of the far-ranging social experiments of Latin America.

In this context, the motivation of the agrarian reform lay in a peculiar set of conditions that made reform part of an ideology which the military junta felt it had a role to carry out. While on the one hand "marginality" and a semi-feudal character to much of Peruvian agriculture were ideological pressures for reform, the final form of the policy is a reflection of the constraints of economic productivity and containment of the newly freed peasantry and rural proletariat. The cooperative in one of its several forms was the solution decided upon by the planners of the agrarian reforms.

To return to the issue alluded to in the introduction to this chapter, whether there was a constant and direct pressure for agrarian reform in the southern sierra, there is considerable doubt. While there was the experience of the land invasions that had swept the sierra in the early 1960s, the movement had been successfully quashed through a combination of police tactics and the failing of the peasant organizer Hugo Blanco; when the 1969 agrarian reform was decreed, there was no viable peasant organizational interest groups successfully pressuring for such a policy. Unlike the Belaúnde agrarian reform, the 1969 decree was not a direct response to pressure from the peasantry and limited to the regions where the pressure was most intense but was conceived as one of a general set of reforms by the military planners. From this point of view, there was a large element of surprise when the agrarian reform was declared and even more surprise when it appeared to be taken seriously by the government bureaucracy.

5

The Cooperative Túpac Amaru II

The Pampa de Anta region has from the beginning been considered a "showcase" for the agrarian reform in highland Peru (John Gitlitz 1971). It was one of the first to be chosen, and much effort was spent in achieving results that the government could use as a model for the indigenous regions of the country. Such a choice was not fortuitous; the Pampa had been declared an agrarian reform zone during President Belaúnde's regime but no land had ever been expropriated. Much of the groundwork was completed during this period. Formal declarations similar to those required under the 1969 law had been obtained from landowners, and a socioeconomic survey (Ministerio de Agricultura 1964), aerial photos, and a cadastral survey were made.

The initial step in the implementation process occurred with the arrival of a team of eight civil engineers and six topographers sent from Lima on 8 August 1969. Because of the work that had already been done and the pressure to produce results, the group was able to complete the bulk of expropriations by December. As Figure 5–1 illustrates, there have been a series of stages of expropriation that are continuing into the present. The bulk of the properties were taken, though, in the first two stages. By the end of this period, the

Figure 5–1. Land Expropriation in the Pampa de Anta, 1971–1973

	Number of Properties	Total Area
First adjudication (June 1971)	65	21,274 has.
Second	9	11,467
Third	4	590
Fourth	8	715
Fifth (August 1973)	19	3,310
Total	105	37,352

Source: Douglas E. Horton 1974:18.1, Table 1–3.

largest estates, including one of 10,666 and another of 2,747 hectares, were turned over to the cooperative. These two estates now occupy 36 percent of the cooperative area (Douglas E. Horton 1974:18.1).

The first wave of expropriations was turned over to a provisional committee, the Comisión de Adjudicacíon Provisional, to administer until a final decision could be made on the method of adjudication. Some of the expropriated properties were farmed during this period, using advisers from the Ministry of Agriculture and peasants who were either recruited to work on the properties or had had contractual relations with the owner prior to expropriation.

Meanwhile, an ad hoc commission was created to study the region, survey the data that had accumulated, and census the population to determine the beneficiaries. It was to be responsible for determining the final mode of distribution. A document (COMACRA 1970) that presents the final report of the commission explains the rationale behind its choice. Two goals were uppermost in the minds of the commission members. First, they sought to select an optimum size economic unit, given the socioeconomic characteristics of the region, that would raise the productivity of the ex-hacienda sector in the regional and national economy. Some of the factors involved in this decision were economy of scale, facility for technical assistance, and the nature of the pre-reform agricultural and cattle operations. As a corollary, provisions were to be included to allow beneficiaries to share in the hoped for increases in productivity. The second goal was to create mechanisms for peasants to participate in the decisionmaking processes, economic and political, from which they were barred in the pre-reform political economy. As the commission perceived the problem, peasants had been marginalized by the concentration of economic and political power in the hands of an agrarian elite. Any change in land distribution would ideally attend to the integration of beneficiaries in a new and more open post-reform political economy.

The task of the commission was not easy. There is considerable evidence to indicate that it was compromised by having to accept decisions handed down from government officials at the national level that often conflicted with ongoing planning and programs in Cuzco and the Pampa de Anta. The basis of the conflict was the political impact of the agrarian reform that the government sought to use in gaining support for its programs at the national level. If the Pampa de Anta was successful as a showcase for widely espoused economic and social reforms in the rural sector of the country, then the government would have a stronger case for garnering support in moving into other agrarian reform zones in the central and southern highlands.

Two examples of major shifts in thinking were the parcellation-collectivization issue, which, one resolved, led to another, the size and scale of the unit selected for the Pampa de Anta. In the first case, there was an initial consensus in creating a small- and medium-sized post-reform tenure pattern with individual holdings (Kuitenbrouwer 1973:13; Harding 1975:235). This formed the basis for the early planning in Cuzco and was communicated to peasants in the Pampa de Anta. Later a collectivization strategy emerged, based on the cooperative model, at the national level that quickly filtered down into offices in Cuzco. This major shift in policy led to problems in the implementation of the agrarian reform in the Pampa de Anta.

The second issue, after collectivization was decided, was the size and scale of the cooperative unit. It was felt by officials in Cuzco, that, given the socioeconomic characteristics of the Pampa de Anta, the most rational plan would be to organize three or four enterprises and a central service cooperative. More enterprises, it was felt, such as one for each of the ten to fifteen naturally interacting sets of communities, would lead to units too small to afford adequately trained advisers. But these plans were scrapped by a political decision at the national level for a post-reform enterprise that would have more impact in the total picture of agrarian reform in Peru (Horton 1974:18.8,XI.4).

The mode of distribution that was selected to meet these ends was a large producer's cooperative in which all the expropriated resources—land, cattle, installations, crops, and equipment—are owned and exploited by the members. No one member received individual rights to exploit any resource. Instead, members provide labor and monetary contributions as the operating capital of the cooperative, although loans have been an important source of capital especially in the early stages of the cooperative. In the period when the cooperative was being planned and members recruited, a loan of seven million soles from the Banco de Fomento Agropecuario was essential in providing an initial input into the operation.[1]

The recruitment of members from among the peasants qualified as beneficiaries of the agrarian reform was crucial to the economic and social success of the cooperative. This task was programmed in a number of ways. The first was a media program, authorized by a provision of the Agrarian Reform Law, and carried out by a special branch of the agrarian reform bureau of the Ministry of Agriculture. There were a number of aspects in the program. First, as required by law, notices of the proceedings were published in the local newspapers, including *El Sol* and *El Comercio* of Cuzco. Supreme decrees signaling key steps were carried in *El Peruano*, the official government newspaper. News reports and stories were given to the press on occasion, but this was infrequent.

Second, an audiovisual branch of the Ministry of Agriculture was created with the explicit purpose of designing a program in which the media would disseminate information concerning the agrarian reform. There were a number of facets to the program. A radio program, the Radio Forum, was created to broadcast news concerning the agrarian reform and the cooperative. The broadcast, in Quechua, began in July 1970 on Tuesdays and Fridays during the dinner hour (7:00 P.M.). Its format was based on the well-known Radio Forum first used in Canada in 1941 and later in India (Paul M. Neurath 1962). A theme was chosen for the broadcast that was then to be followed by a discussion among the assembled peasants. A secretary was elected to take notes on the discussion; any points not cleared up were referred by him to the appropriate branch of the Ministry of Agriculture. Another facet of the media program was news involving the cooperative broadcast in Quechua from a van equipped with a public address system that crisscrossed the Pampa. And finally, one of the most attractive media forms were well-done posters displaying slogans of the agrarian reform and quotations from President Velasco. For example, a widely disseminated poster compared the role of the agrarian judge with the notorious *tintorillos,* or mestizos who had made a living through forging and otherwise falsifying land titles and other documents used

1. Reported in *El Comercio,* 30 March 1971.

to exploit peasants. The posters soon found their ways onto walls of stores and peasant homes next to the ubiquitous *almanaque,* calendars depicting days reserved for patron saints, and pictures of Presidents Belaúnde and Velasco.

Besides the media campaign, interpersonal communication processes were utilized to prepare the peasant beneficiaries. The basic process is referred to as *concientización,* consciousness raising. It involves the contact between a representative of the Ministry of Agriculture, a *promotor,* trained in the objectives of the agrarian reform, and the peasant community. The *promotor,* who may travel with one or more colleagues, visits the community, asks to be given a room to stay in during his presence, and explains to the authorities that he is representing the Ministry of Agriculture and wishes to hold a series of talks with the assembled peasants. The object of the talks is to explain the formation of the cooperative, the goals of the agrarian reform, and to motivate the beneficiaries to join the cooperative. Peasants are then asked to send a group of representatives to a seminar[2] held in a regional urban center where the cooperative scheme is explained in detail. The representatives return to the community and explain the scheme to the assembled peasants.

In July of 1970 meetings began between the *promotores* and the peasants, and the cooperative mode of land distribution was first presented. Two representatives from each community were elected and sent in late October of that year to an eight-day seminar held at Sullupucyo, the large hacienda previously owned by the Luna family and now the cooperative headquarters. Other seminars were held in Cuzco in November and in Yucay in June of the following year; they were attended by community representatives. After each seminar, representatives returned to their communities and reported on what happened in the assembly. After the cooperative was organized and underway more specialized seminars were held such as one for cooperative administrators in Sullupucyo. The initial seminar in October of 1970 was conducted by specialists from the Cuzco offices of the Ministry of Agriculture; after that

2. The seminar, or *cursillo,* has been one of the basic devices employed by agrarian reform institutions to effect an exchange of information. It has been used on all levels: for employees of the agrarian reform bureaucracy, for peasant leaders, for members of the Guardia Civil, for *promotores,* and for peasant beneficiaries.

date the Centro Nacional de Capacitación y Investigación para la Reforma Agraria (CENCIRA), headquartered in Lima, began to take part in running the seminars.

Formal Organization of the Cooperative

The cooperative is an economic unit of production in which members have a voice in the setting of social and economic policy. On the one hand, members are a source of labor exploited by the cooperative; on the other hand, there are mechanisms by which members can articulate opinions and gain information concerning cooperative policy.

The most important decisionmaking body is the general assembly. It is made up of 120 delegates elected from among all of the members to represent different geographical units in the region bounded by the cooperative. The general assembly ordinarily meets twice a year, and extraordinarily as the occasion demands, to decide on the major issues and policy confronted by the cooperative. All of the major financial and administrative policies are decided by this body, based on a simple majority of the delegates present at a meeting.

The general assembly is responsible for an annual work plan that sets long-term policy and priorities. The allocation and exploitation of cooperative capital falls within the domain of the annual work plan; one important initial set of decisions faced by the embryonic cooperative was, for example, the percentage of cultivable land devoted to potatoes as opposed to pasture.

The general assembly does not actually carry out the policy it sets. This task is the responsibility of the administrative council made up of eleven elected members that meets at least once a month. This body operationalizes, on a short-term basis, the long-term decisions made in the general assembly. It acts as the agent in contracting temporary and permanent personnel, lets contracts, calls general assemblies, constructs an operating budget, and performs a number of other activities outlined in Article 67 of the Statutes of the Cooperative.

One of the most important functions of the administrative council is to delegate the actual day-to-day administrative duties to an administrator who is contracted from outside the cooperative and responsible to the administrative council. The administrator oversees all of the technical operations of the cooperative in the Pampa de Anta region. He has the final word concerning a specific field procedure, such as the selection of seed potatoes, storage practices, and insect control. He has considerable power and is the individual most responsible for the success or failure of the annual work plan. Although he works closely with the administrative council, he is not directly contracted by it. The council submits a list of three names to the Agrarian Reform Section of the Ministry of Agriculture, which selects one of the three.

The functions of the administrator changed somewhat following a reorganization of the cooperative in January 1972. This change is illustrated through a comparison of two diagrams of cooperative organization (Figures 5-2 and 5-3); the first diagram represents the initial organization of the cooperative, and the second represents changes made that went into effect in January 1972.

In the first diagram, the responsibilities of the administrator (*gerente*) are clearly delineated: he is charged with administering various divisions of the cooperative, such as the accounting division, the general services division, the Chamancalla cattle operations, and the Ancachuro agricultural operation.

The second diagram reflects changes made after the first period of operation. The most significant change was the substitution of an executive committee for the administrator, a result of the complaints that arose during November and December of 1971. There was then a widespread dissatisfaction with the administrator, particularly since he received a salary of thirty thousand soles that seemed exorbitant in relation to the daily wage of twenty-four soles received by an ordinary member. Additional complaints concerned the amount of work put in by this individual. The executive committee was created to resolve the problem. The committee was composed of members elected from the delegate assembly and was placed under the permanent "legal" assistance of the government cooperative agency.

In addition to the administrative organization, until 1972 there was an organization for labor recruitment within the cooperative. Labor committees were located at the district level and were charged with recruiting labor as needed from among cooperative members. By working with a labor committee attached to the administrative council, they would act

Figure 5–2. Cooperative Organization, 1971

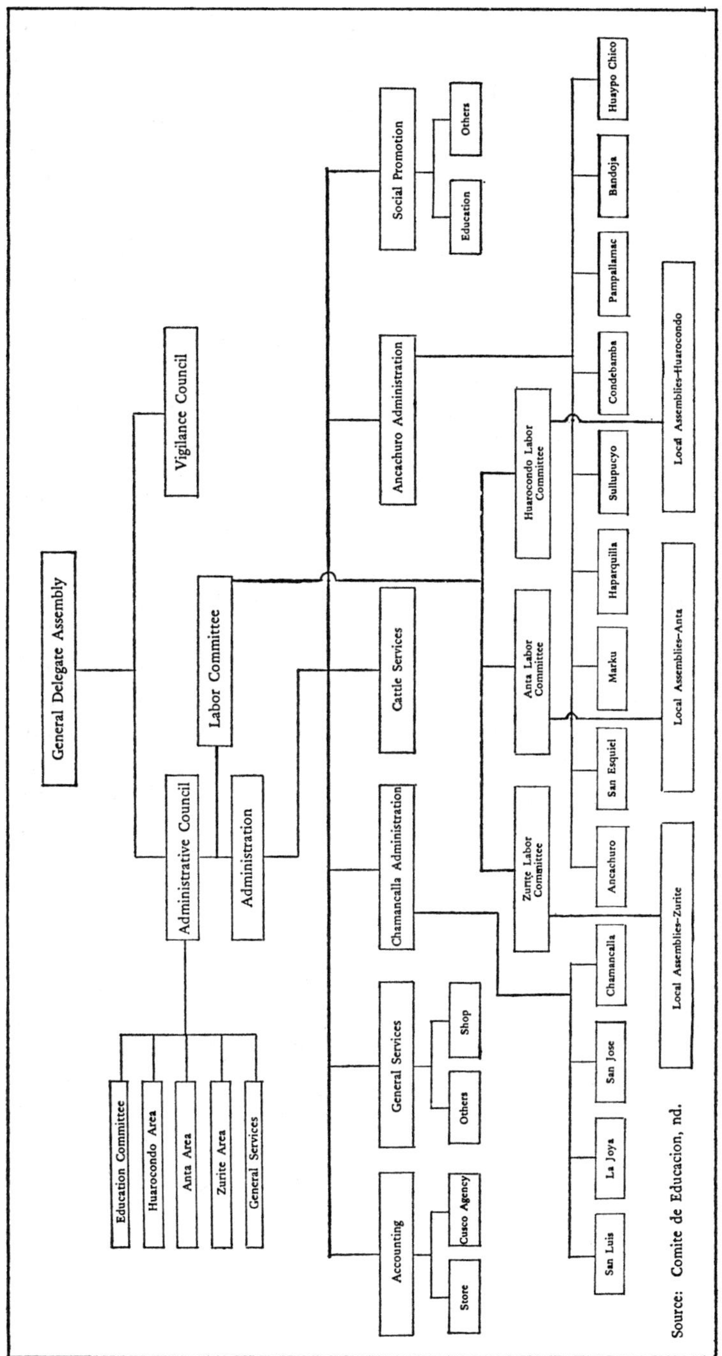

Source: Comite de Educacion, nd.

on directives to ensure a continual and stable labor supply. Usually a local assembly would be located wherever a nucleus of cooperative members was to be found, often among an ex-hacienda population or within a larger independent peasant community. Rumipata, Antapampa, and Tukiwasi each had a local assembly of cooperative members. Local assemblies elect delegates who represent them in the general delegate assembly; delegates are the main link between the individual member and the upper echelon of the cooperative administration. Local assemblies are quite important and are perceived as the loci of participation of members in the policymaking process of the cooperative.

The shift from local assemblies prior to 1972 to production units in 1972, reflected in Figure 5–3, resulted from a lack of viability of the local assemblies, which will be discussed later in this chapter, and their consolidation into larger geographical units. One result of this change was a bipartite division of the cooperative into a cattle operation and the agricultural operations that were somewhat intermingled before. Additionally, there were a set of offices created at the local level of the production unit: the *jefe, sub-jefe, mayoral,* and *caporal.* The functions of the labor organization network were incorporated into the production units. These changes that went into effect in January 1972 appear to have resulted in a more streamlined operation with more power vested in fewer individuals at the local level and a more direct connection between government agencies and the activities of the cooperative.

The organization of the cooperative may now be summarized by distinguishing three levels of organization: an upper level of administrators consisting of elected officials and advisers from government agencies; a second level composed of individuals involved in administering the day-to-day operations of the cooperative (prior to 1972, this level would include a contracted administrator and the heads of the various local assemblies, later superceded by an executive committee and the officials of the local production units); and a third level consisting of the individual cooperative members. Delegates functioned to link each organizational level, to transmit information, and to act as a feedback mechanism to enable individual members to provide input into the long-term policies of the cooperative.

Figure 5–3. Cooperative Organization, January 1972

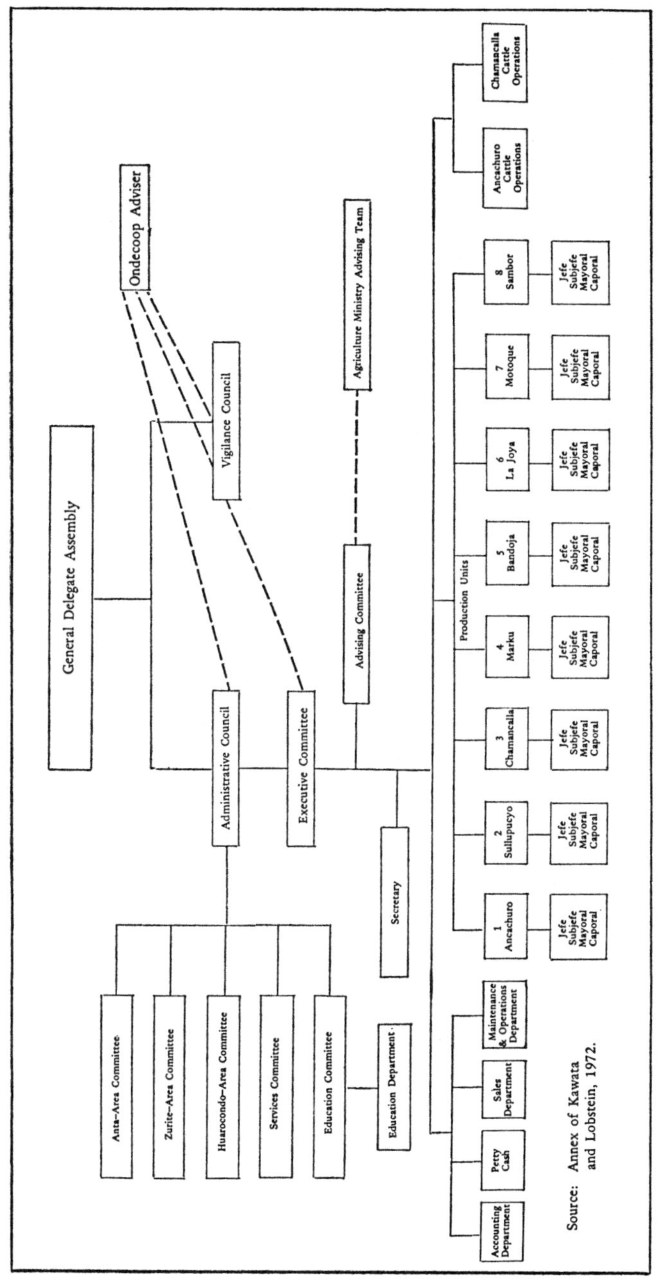

Source: Annex of Kawata and Lobstein, 1972.

One important aspect of this multilevel organization is that individual members have no direct voice in the short-term production decisions. These decisions are the domain of the administrator (pre–1971) and his advisers and representatives involved in the operations of the various ex-hacienda properties. Members do have an input into the setting of the overall long-term policy of the cooperative, but this input is through the voice and vote of their elected representatives or delegates. Once the annual work plan is set up, it is out of the hands of the individual members to revise or oversee.

The Peasant Communities Law

During the early stages of the agrarian reform, representatives contacted peasant communities to begin to implement the provisions of the Peasant Communities Law enacted in 1970 as part of the overall agrarian reform program. In the independent communities elections were held late in 1970 according to the new slate of political offices stipulated in the law. No other major provisions of the law were carried out, however, such as the restructuring of the land tenure pattern of each community on a cooperatively owned basis. In the hacienda communities scheduled for expropriation, peasants were given the option of petitioning for formal recognition as a peasant community, or forming a communal cooperative according to a new law regulating the production cooperatives organized as part of the agrarian reform and a new type of cooperative, the communal cooperative.

Changes in the Regional Political Economy

The direct effect of the expropriations was to remove hacendado control over the political economy of the region. Virtually all of the major haciendas were expropriated in the first two stages; by the end of this period there was no effective hacendado presence to be found in the Pampa. Those hacendados who had maintained a residence there at the time of expropriation moved to Cuzco. Some bought properties in the "safer" regions such as the *selva*. Others gave up agriculture altogether, went into retirement, or pursued occupations in Cuzco, or the other sierra urban centers. Many migrated to Lima to live. In many ways, agrarian reform brought to culmination a trend of the elite away from an agrarian lifestyle based on the power and prestige of land. As the larger units became broken up through inheritance, the heirs preferred to maintain a more "urban" lifestyle by moving to Lima, seeking a career in civil service or in private enterprise, and disassociating themselves from a rather quaint lifestyle based on the hacienda system.

In part such a trend explains the lack of a viable opposition by the hacendados to the implementation of the agrarian reform. There was little outcry from the traditional spokesmen of the hacendado elite, that is, the newspapers and the regional landowners association. Most of the attempts to create obstacles occurred at the lower rungs of the agrarian reform administration: in particular, hacendados tried to divest themselves of *feudatarios,* sold cattle and equipment, contracted to have their eucalyptus trees cut and sold, and, in general, tried their best to liquidate their holdings before they became subject to expropriation. In a few cases there were conflicts of interest among administrators employed in the agrarian reform bureaucracy; some were themselves property owners, and many were friends of hacendados who were involved in expropriation proceedings. At any rate, the lack of an organized and viable opposition at the departmental level is in part an indication of the real extent to which the sierra hacienda sector of the regional and national economy had weakened over the years.

Certainly actions taken by the military government to ensure that interest groups at the regional level could not create obstacles were strategic. These actions were primarily institutional transformations to shift power to the national level through an expansion of the bureaucracy, the replacement of middle- and upper-level bureaucrats by individuals brought into the bureaucracy from outside the region, and a process of centralization of decisionmaking that is continuing to the present.

An initial expansion was made during the expropriation phase as teams of topographers, civil engineers, and other technicians arrived from Lima to increase the capacity of the regional office of the Ministry of Agriculture to process the expropriation proceedings. Subsequently, all of the major administrative branches of the government dealing with the rural sector, particularly the cooperative agency, the agrarian reform agency of the Ministry of Agriculture, and community development agency, and the peasant communities agency, were expanded to some degree or another. This created problems with obtaining sufficient personnel to fill the new positions. Lower administrative positions were easily filled from the local labor pool since they were based on pay scales much higher than the private sector could provide. There was more difficulty in filling the skilled and professional positions at the middle and higher rungs of the government bureaucracy. This led, in many cases, to the transfer to Cuzco of personnel working in regions where the agrarian reform was yet to enter. Many of these individuals were graduates of the coastal universities, in particular Arequipa and Lima.[3] Often these individuals were not familiar with the peculiar characteristics of the southern sierra region. This was not seen as problematic though; on the contrary, it was advantageous to bring in personnel from outside the department as a means to eliminate potential conflict of interest situations that might occur.

There was considerable activity in the departmental capital following bureaucratic expansion. Shiny new Dodge pickups and mobile vans began to be seen on the streets of Cuzco, causing more than one lifelong Cuzco resident to comment on the invasion of Cuzco by *criollos* from the coast. This activity carried over to the pampa where during some days three or four vehicles of different ministries could be found in Izcuchaca.

Along with the expansion of the existing bureaucratic network, the momentum of the implementation was rapidly increased. There was widespread awareness that this was necessary to keep the agrarian reform from stumbling into stupor.[4] This awareness was translated into pressure from the national level to move the implementation along. Numerous visits were made by various high-ranking officials from Lima, and President Velasco himself visited the southern sierra in a well-received trip during September 1971.

Another move by the military government was to create an agrarian tribunal and land judge, *juez de tierras*, attached to

3. The fact that many of the military reformers came from Arequipa gave individuals from that city special status as potential government employees.

4. This was the major recommendation of an early evaluation of the agrarian reform by the economists Edmundo Flores and Solomon Eckstein (*El Trimestre Economico* 38:3 [1970]).

the Ministry of Agriculture, which was specifically charged with ruling on conflicts involving agrarian law. Virtually every major type of problem involving peasant communities, including recovery of alienated communal lands, demarcation of boundaries, dismissal of tenants, and the payment of rent, was subject to arbitration by the agrarian tribunal. Peasant litigants were exonerated from paying any costs involved in litigation. The creation of such an independent judiciary was a major tool in removing control and influence over land litigation involving peasants that hacendados and their representatives had exercised. The first judge arrived in Cuzco in September 1969 and was followed by another in December. They soon were at work both in Cuzco and in numerous visits to the localities where litigation was in process. Their Volkswagens with the seal of their office were familiar sights along with the pickup trucks of the other government bureaucracies.

To control more effectively the implementation of the various reforms and to hold in rein a potentially inefficient, expanding bureaucracy, a further attempt at institutionalizing the reform occurred on 22 June 1971, when an umbrella-type organization, the Sistema Nacional de Apoyo a la Mobilizacion Social, SINAMOS, was created by presidential decree. The stated purpose was to assist in the satisfaction of the goals and needs of the popular classes and to enable power to make decisions affecting one's life to be transferred to the people. In practice, it appears to be an attempt to control the impact of the revolutionary changes set loose by the government and to guide the course of spontaneous change in directions that conformed to national level policy.

Although SINAMOS worked closely with other national agencies in the implementation of the agrarian reform, it first surfaced in the sierra as the agency responsible for the mobilization of peasant support for President Velasco during his trip to the sierra in September of 1971. SINAMOS officials were conspicuous in the presidential party and were generally given credit for organizing the massive turnout of peasants to see Velasco during his stops at Sullupucyo, the ex-hacienda headquarters of the embryonic cooperative in the Pampa de Anta, and at other locations in Cuzco and Puno Departments.

Following Velasco's trip, SINAMOS returned to the reorganization of the various state agencies that were to be incorporated into the umbrella organization under the June decree.

These agencies included the cooperative agency; the community development agency; and the agencies concerned with peasant organizations, peasant communities, and the agrarian reform, among others. Personnel employed by these agencies were to pass over to SINAMOS and a new General Law, *La Ley Organica*, would be issued to set guidelines for its operation. The new law appeared in April 1971 by supreme decree. In it, SINAMOS is a cabinet level agency with a director reporting directly to the president; the director is in the same category as a minister and is allowed a vote in the Ministry Council. Its structure is made up of four levels of administration corresponding to administrative levels of the urban hierarchy. These include the office of the director in Lima, regional offices located in the macro-ecological zones of the country, zonal offices, and lastly, local teams operating out of the zonal offices directly with popular sectors. All funding for projects previously handled by the separate agencies was now to be channeled through SINAMOS: similarly, all policy decisions were to originate at some point in that organization.

With SINAMOS the process of consolidation and centralization of decisionmaking in a bureaucracy that was directly linked to the control of the military government was complete. While it seems clear that many of the programs concerned with the popular sectors now could be incorporated into an overall development strategy of the State, on the other hand, the power of the State was strengthened in its ability to control the thrust of the popular sectors. Many of the reforms removed the constraints that had controlled the thrust of the popular sectors; while SINAMOS was designed to channel these forces into what was felt to be best by the national government, the possibility still existed that only those avenues of change that did not threaten the stance or policy of the State would be permitted under the new rules. At any rate, SINAMOS placed the reforms of the military government into a perspective that both strengthened the government in administering the reforms and in controlling threats to them.

The net effect of the first phase of the agrarian reform to this point was to radically alter the constraints under which the political economy of the Department of Cuzco, and, in extension, the Pampa de Anta, operated. The expropriation of Pampa haciendas was accomplished by the replacement of the hacendado elite's control over the regional juridico-political institutions with an agrarian reform bureaucracy expanded, reinforced, and manned with personnel independent enough and willing to carry out reform directives emanating from Lima. The regional political economy, in effect, was helpless to offset the shift in power to the national level, and, inasmuch as the political economy of the department was closely correlated with mestizo control over crucial institutions of the Pampa de Anta, similar changes occurred there. Whereas before, reform policies that threatened mestizo interests would have been thwarted, mestizos now could no longer rely on the support of the agrarian elite.

The Communications Program

It would appear that the communications program designed by the government would have been ideal: it combined the two pillars of the model of the diffusion of innovations, mass media, and interpersonal communications, in a well-conceived and sincere effort in reaching out to the beneficiaries. Nevertheless, severe obstacles, to which we will now turn, constrained the program at the national and regional levels.

First, the content of agrarian reform and cooperative policy is extremely difficult for the average peasant to comprehend. All relevant legislation and policy including the agrarian reform and its annexes, the Peasant Communities Law, the General Law of Cooperatives, and the rules and regulations of the Túpac Amaru II Cooperative are in Spanish in a region where ordinary discourse is carried on in Quechua. According to the 1970 Peasant Communities Census, 45.7 percent of the adult population had no schooling whatsoever, and 33.8 percent of the remainder had not completed *primaria* the lowest grade of schooling. In order to explain the most basic workings of the cooperative, sophisticated Spanish concepts must be used with reference to legislation as the ultimate source of definition. For example, in a pamphlet (Comité de Educación, n.d.) written by advisers to the cooperative and disseminated to peasants throughout the pampa, one section details the requirements for becoming a member of the cooperative.

> In order to attain membership standing, the Constitution of the Cooperative requires that interested persons:

a. ask the Organizing Committee for membership and sign the respective list; previous approval by the Agrarian Reform Agency is needed.
b. not have conflicting interests nor be in commercial competition with the Cooperative.
c. not belong to another Cooperative of the same type of activity nor belong to an organization whose purpose is incompatible with those of the Cooperative.
d. the feudatarios must contribute their part of their indemnization as provided by article 184 of the official text of law 17716 as their share (in the Cooperative).
e. comuneros must be approved by the Peasant Communities Agency and their community must be a member of the Cooperative.
f. the entrance fee must be paid and at least one share (certificado de aportacion) must be purchased. [Article 8 of the Statutes of the Cooperative Túpac Amaru II of Antapampa].

Persons who will be able to seek membership under the Constitution of the Cooperative are provided for in Article 7 of the Statutes and who besides meet the following requirements:
a. they must not be disqualified by paragraphs "b" and "c" of article 8 of the Statutes. At the same time, feudatarios and comuneros must meet the requirements of paragraphs "d" and "e" of the same article.
b. they must complete an application directed to the President of the Administrative Council.
c. they must pay the entrance fee and purchase at least one share and contribute a sum of 100 soles.
d. they must be accepted as a member by resolution of the Administrative Council in agreement with the procedure described in Labor Legislation [Article 9 of the Statutes]. (Comité de Educación n.d.:4–5.)

The reference in the text to the Statutes of the Cooperative involves a complex legal document, and, like all such legislation, requires access to reading and communicative skills far beyond the range of most peasant speakers of Spanish. A further example of the impass legislation presents is the labor legislation passed by the delegate assembly of the cooperative, which contains over 309 separate provisions governing the obligations and responsibilities of cooperative members. Admittedly, the labor code is a model of the most contemporary and sophisticated thinking regarding labor legislation, yet rules and regulations like the labor code are the base upon which the cooperative operates and the member must know and abide by them. To the peasant contemplating joining the cooperative, the maze of rules and regulations caused considerable uncertainty and anxiety about the uses to which such legislation could be put.

Peasants were well aware of the density of the legislation and their inability to digest it. One solution was to operate in Quechua. This solution called for a translation of the law into Quechua in May of 1970 before the actual creation of the co-op.[5] The pressure was a reflection of the fact that, as in the earlier abuses of the *tintorillos,* manipulation of legal loopholes and procedures was a common means for personal aggrandizement. Already teachers were found to be manipulating the law on the local level by cultivating plots owned by *comuneros,* registering themselves as *feudatarios,* and drawing up papers with the assistance of lawyers documenting their possession of the land.[6] In many cases, individuals accused of misdeeds were residents of the community at one time, had gone through the educational system, and joined state service as teachers, lawyers, or other officials, but continued to maintain ties with the community. By law, though, they were prohibited from registering as *comuneros.*

Translation of the Agrarian Reform Law, and the other important legislation, into Quechua would not have solved the problem. Quechua is not a written language, and although orthographics do exist, it would require reading and writing skills that the peasants do not possess since the Quechua-speaking peasants are illiterate in their native language. Also, direct translation of the law into Quechua would have made even more difficult a correct interpretation of a law based on Spanish legal concepts and phraseology. Yet, it certainly would have strengthened the role of intermediary groups who would articulate the law to peasant beneficiaries.

A second problem area lies in the formulation of major policy decisions. Often they are made at the national level, without adequate consideration of their effects on the communication process at the regional and local level. The following example is one of the most blatant, coming at a time when communication of the intent of the agrarian reform was crucial. During the initial seminars held between representatives of peasant communities in the pampa and advisers of the

5. Reported in *El Comercio,* 6 May 1970.
6. Reported in *El Comercio,* 30 September 1969.

various agencies implementing the agrarian reform, peasants were told that the distribution process would involve divisions of expropriated land into parcels of three hectares per household. The plots would be farmed cooperatively by labor groups like *ayne* exchange labor groups. The cooperative would assist by providing additional labor for peak labor periods and function as a service center dispensing fertilizer, insecticides, and other capital inputs at favorable prices.

Further, there would be a large amount of pasture land set aside that would not be parceled. There was considerable enthusiasm for this plan but later, when it was revealed that parcellation would not occur but instead expropriated land would be distributed to a cooperative, disenchantment set in. This change was a result of a reorientation of agrarian reform policy at the national level away from earlier thinking when there was a clear orientation toward a small and medium size post-reform tenure pattern (Kuitenbrouwer 1973:13). Its effect in the pampa was to create confusion, skepticism, and accusations of "liar" placed on the original representatives who attended the seminars.

Later, after the cooperative was organized and functioning, major policy decisions continued to reflect disproportionate input from advisers and technicians whose primary allegiance was to the national level. Members with valid complaints were not recognized either collectively through their delegates in the delegate assembly or individually in local assemblies. The lack of input and feedback is articulated in an interesting document (Na. 1971) written by a group of seventy-eight delegates from the local committees and six elected officials of the cooperative. According to the document, a major problem was the lack of information on the activities of the cooperative. Such information, it implies, was a domain of the advisers, technicians, and representatives of the national agencies. Out of a total of eighty-four officials who signed the document, "only five know what is the total of the financial loss incurred in the last year, six know what the functions are of the local assemblies, six know what the functions are of the Education Committee, and one knows what the Exploitation Plan is" (Na. 1971:3). The statement continues: "Knowledge comes from being informed, not from having participated in these activities. . .if they tell us we are going to buy tractors, agricultural implements, perhaps we would say yes, because we do not know what the cost will mean, how it is going to be paid, nor how many fewer days of work it is going to represent for us" (p. 2).

The document continues with a plan constructed to solve these problems:

> *How we Should Prepare Ourselves*
>
> Many functionaries and promoters have come before us who have told us many things and made us many promises, and almost all have been lies. We no longer want them to teach us what happens to strike their fancy.
>
> We don't want promoters who bring us their ideas or the ideas of the functionaries; we want them to help us clarify our own ideas and be able, together, to arrive at a single idea for which we will work.
>
> Now we want them to speak to us in Quechua; we want to discuss with them what they are going to teach us and tell them that the problems of the individual are to be considered the problems of everyone. We want them to teach us not by talking, but by doing the chores related to our work, in our Cooperative: not that they do the chores, but that they aid us in discovering the different ways of doing them, the pros and cons of each way, but that we choose and do what we consider best for ourselves. They will be advisors.
>
> We want our sons, brothers or some of ourselves, the present leaders (*dirigentes*) to be prepared by the technicians who will come, in order that we who live in the community may prepare the rest of our brother and sister peasants.
>
> The technicians that come must have a social consciousness and besides, know how to guide a plow and to seed potatoes. If they can do this then they will know how to enable us to:
>
> have new ideas
> gain greater knowledge in order to have a better life
> know how to discover the real necessities
> learn to read and write
> decide responsibly
> be financially honest
> organize ourselves well
>
> We also want the technicians to be called in accordance with the Quechua term "Yachay Mastarig" which means he who provides ideas. And, we want our brothers who help prepare the rest of us to be called "Wayqeichis Rikcharinchi" which means he who makes others awaken.

This document articulates the friction that occurred during the first year between the advisers and technicians and the upper- and middle-level officials elected from among the membership at large. The latter felt that important decisions were made by the advisers and technicians in conjunction with some easily persuaded elected officials. These criticisms were well founded. Because of the importance of the cooperative in the national agrarian reform program there was a tendency to watch closely the trajectory of the cooperative to forestall any failure that might occur. One disaster had already occurred: The Alto Urubamba Cooperative located in the politically unstable Convención Valley was taken over by the government in February 1972 after showing signs of mismanagement and a split between middle- and upper-level administrators. This cooperative was the other showcase in the sierra designed to exemplify the goals of the agrarian reform. During the same month there were rumors that the Túpac Amaru II Cooperative in the pampa was destined for the same fate. While such drastic actions may have been necessitated by irregularities in the organization and functioning of the cooperatives, there is considerable evidence that government agencies overestimated disrupting tendencies occurring normally in a rather radical transition in the political economy. The tension between overly cautious centralized control by the government and a local autonomy essential for the cooperative to work has been a core problem and a dilemma begging resolution.

The creation of a formal communications program designed to reach out to the beneficiaries of the agrarian reform was combined with an effort on a different front to link the national and regional levels with individuals who were more intimately in contact with peasants. I have pointed out that one of the major effects of the agrarian reform was to remove the basis of the power of the mestizos in the regional political economy through expropriation of hacienda lands and their replacement with an agrarian bureaucracy operated out of Lima. Mestizos were bypassed by government agents because they represented a class grouping opposed to the agrarian reform. Yet it was obvious to the planners of the agrarian reform that such intermediaries had been instrumental in communicating information to peasants before the reform, and in order to be effective the bureaucracy would have to use similar intermediaries.

The problem was resolved in two ways: First, in some instances a compromise was reached on the local level by allowing mestizos to continue in key roles where the situation so demanded. An example of such a compromise will be presented in the next two chapters dealing with events at the local level. Second, at the regional level, representatives of the agrarian reform established links to the peasantry through individuals who were perceived to correspond to the "peasant leader" image. This role is based on the archetype of Hugo Blanco and other lesser figures who were instrumental in mobilizing the peasant mass during the uprisings and land invasions of the early 1960s.

When it became obvious that linkages needed to be established to the peasantry, the peasant leader appeared to be the solution. Unlike mestizos, the traditional intermediary group, the peasant leader appeared to articulate the interests of the peasants and to be ideologically opposed to the hacendado class, which the agrarian reform was designed to remove.

The intermediary niche was also sought from below: It represented a source of power for political entrepreneurs interested in maximizing their position in the new political economy. A struggle developed among those who sought to appear as peasant leaders and reap the rewards to be had. Many lost out in the competition that followed; others like the example of EP were successful and achieved legitimacy through recognition by representatives of the agrarian reform bureaucracy.

When EP appeared during the initial contracts between bureaucrats and peasants in early 1970, he claimed to be a spokesman for all the peasants in the Pampa de Anta. His credentials supported his claim: he lived on a large hacienda in the pampa where he had represented *feudatarios* in a struggle with the owner for some years; he was educated in agronomy at the University of San Antonio Abad in Cuzco and received ideological legitimacy through study at the Patricio Lumumba University in the Soviet Union. Besides these credentials, he was adept at using threats of mass mobilization of peasants if his position was not accepted. These threats were taken seriously, and, out of fear as much as a real understanding of his following, he became accepted as the major spokesman for peasant interests. EP has been in on the ground floor ever since his initial attendance at the first planning seminar held for peasant representatives in Arequipa in 1970. His position was solidified when President Velasco singled him out in a group of leaders that met with him in his trip to Anta in September 1971. He has played a major role in policymaking in the cooperative and has considered the peasants in the pampa regarding every major initiative by governmental agencies.

While EP enjoyed support and legitimacy from above, he has not commanded a following from below. Outside the hacienda in which he resides, he is perceived by peasants as a self-interested *misti*, totally opposed to peasant interests. In some instances he has even created havoc that was later smoothed over by government action. In late February 1972 he broke into the home of an aged widow with a group of his followers and stole some items, on the excuse that she was speaking out against the agrarian reform. He was immediately put into jail by the Guardia Civil but later released by order of SINAMOS. Peasants were outraged at the events and in a meeting in Anta denounced him.

The effect of granting legitimacy to EP is to suggest to the peasants that the policymakers are not aware of EP's real standing either through fault or design. Instead of bringing stability to an unstable situation caused by changes in the political economy, EP's role further increased the instability and in extension the uncertainty about the future of the cooperative.

In sum, the communications program was constrained at the regional level by the nature of the program it was attempting to disseminate, the effect of centralization of policy decisions, and the nature of the links established by the government to replace those associated with the pre-reform political economy. In the first case, the complexity of the legislation and the use of Spanish as a medium hindered its communication to an indigenous peasantry that continued to remember the abuses to which such legislation had been put in the past. The effects of centralization were to increase the uncertainty: The geographic and social distance between planners and beneficiaries inhibited the feedback necessary for policymaking; policy shifts were abrupt and reflected lack of consideration of their effects on the communications process; and, lower and middle rungs of the cooperative hierarchy were often unclear about policy decisions made at the upper levels. And finally, "peasant leaders" legitimized by governmental action were often political entrepreneurs who were self-interested and did not represent their peasant constituency. In retrospect, a number of constraints seriously hampered the communications program. As we will now see in the next two chapters, peasants resorted to traditional sources of information at the local level in order to supplement formal sources in arriving at a decision regarding the cooperative.

6

Peasant Decisionmaking: The Economic Context

One way in which peasants could obtain information concerning the agrarian reform was to observe the activities of the Provisional Administrative Commission (PAC) and, later, the cooperative, at the local level. PAC was provided by a clause in the agrarian reform law, in order to administer land selected for expropriation until it could be turned over to beneficiaries in the adjudication process. It consisted of a president, secretary, treasurer, and two assistants, all of whom were peasants, and a group of advisers employed by the Ministry of Agriculture. The advisory group included a veterinarian, a livestock specialist, two agricultural engineers, an agrarian specialist, and various others who assisted from time to time.

The activities of PAC gave peasants the opportunity to observe the economic activities of the then embryonic cooperative. They involved decisions concerning the exploitation of local haciendas and thus brought the agrarian reform down to the local level. This was extremely important in the sense that peasants perceived the agrarian reform, at least in the initial stages, in terms of how it affected the local haciendas that they were familiar with. In each community there were varying degrees of relations with adjoining haciendas; these relations can be described as a continuum with two poles represented by Tukiwasi on one end and Rumipata on the other. In Tukiwasi, virtually all peasants were *feudatarios* residing on the hacienda and completely subordinated to its control. Rumipata, on the other hand, was relatively free of hacienda influence; although there was a hacienda adjoining the community, only a few peasants maintained *yerbajero* relations with it in order to gain access to its fertile pasture lands. "In between" was Antapampa, with a mix of *yerbajeros* and *feudatarios* who lived in the community and worked on the adjoining hacienda.

In each community, peasants were acutely aware of the presence of the hacienda. They knew that at some point in time the land upon which the hacienda lay belonged to the communities and had been alienated from their control through manipulation of the juridic-political institutions. They had always maintained a defensive posture toward the hacienda and often had had to resort to negotiations and/or violence to defend their boundaries. Thus, when the agrarian reform was decreed and expropriation rumors filled the air it was only natural that peasants would look to the local hacienda to see how it would be affected by the reform.

What peasants looked for in the PAC activities and later in the cooperative were based on two concerns: First, they were interested in activities that would bear on their motives, real or imagined, for joining the cooperative then in the process of formation. In some cases, peasants who had maintained relations with the adjoining hacienda as *feudatario*, *yerbajero*, or wage laborer were anxious to see how PAC and the cooperative would change those relations. In other cases, peasants were persuaded by the communications program that there might be benefits to joining the cooperative, such as a secure source of high-paying wage labor, an investment scheme with a return on one's investment, and somewhat later, the availability of high-quality pasture land. They were concerned with how PAC and the cooperative would facilitate and deliver on these promises. Second, as it became clear that all the haciendas would be brought into one large producer's cooperative, peasants began to look at the cooperative as a long-term production unit. Some peasants who had a better understanding than others obviously looked for different things than those who did not understand the nature of the cooperative. But certain activities could be assimilated by any peasant and certainly had a relation to the cooperative as an economic unit. These included the production and distribution functions of the cooperative: the type, origin, and administration of production decisions in the agricultural and cattle operations, and the amount, quality, return, and sale of the harvest produced on the lands. They were acutely aware that PAC had assumed a large loan in order to provide operating capital during the transition period; although advisers consistently denied that the cooperative was responsible for PAC's obligations, peasants believed that the loan had to be paid off by the sale of harvests of the cultivated lands. This loan, plus the debt incurred in paying the owners their calculated indemnization, was felt to be a major fiscal problem incurred by the cooperative.

Rumipata

Rumipata prior to the agrarian reform was not highly involved in the activities of the adjoining hacienda. It was a small hacienda, run essentially as a cattle operation, with only a few *yerbajeros* from the community pasturing their cattle on its relatively fertile pasture lands. Although expropriation was completed in June 1971, PAC began to administer the hacienda during the 1970–1971 growing season. The first thing that the commission did was to plow up the land and seed potatoes. At first glance, this was a rational short-term decision; yields should have been high since the land had been fallow for a long time and the fertility increased by the droppings of the cattle. Using labor recruited by PAC from local peasant communities, the land was prepared, the crop seeded, a series of soil depth adjustments and weedings carried out, and the crop eventually harvested. Considerable capital inputs, such as fertilizers, insecticides, and mechanized technology, was applied.

It is difficult to get a clear picture of the amount of potatoes that were havested from the fields of the ex-hacienda, but the large amounts of capital input and high soil fertility were certainly factors that helped to ensure at least an average output, reported climatic vagaries notwithstanding. Informants are clear in singling out a series of impressions that were common to all of the peasants who observed the 1970–1971 operation.

First, it was clear from the manner in which the land was exploited that the technical decisions were being made by PAC advisers rather than the peasant members of the commission. The advisers visited the ex-hacienda, consulted with peasants, and made the ultimate decisions on a day-to-day basis as to the agricultural practices. They were new faces to the Rumipatiños, recruited in Cuzco or in other cities for the expanding reform bureaucracy, and often fresh from training. There were some faces that were known to the peasants; for example, one of the first local level administrators, called the *mayordomo*, actually had been *a mayordomo* of the Hacienda Sullupucyo owned by the most exploitative and notorious landowner in the pampa.

To the peasants, the advisers initially were an unknown quantity that as the growing season progressed demonstrated a lack of awareness of cultivation practices, as to when the first soil depth adjustment was made, or, when quinoa was seeded, and, in short, all of the questions that were necessary to establish the range and variation in local agricultural practice. This was an indication to the peasants that, in effect, the technicians did not know what they were doing. If, as their title made claim to, they were experts in agriculture, then it was incongruous to the peasants that they should have to ask such basic questions.

Their suspicion was heightened by the fact that the advisers were essentially "urban" mestizos who were new to the community and often residents of regions outside the south sierra. Many subtle forms of paternalistic behavior continued to filter communications between the technicians and the peasants. There was a definite lack of communication in this period that was mutually reinforcing; that is, a patronizing attitude on the part of the technicians was met by a submissive and closed response from the peasants. Far from being a general behavioral pattern, such behavior was a defense against what was perceived as potential exploitation from "urban" mestizos.

Some of the practices of the technicians justified the assessment of the technicians in the eyes of the peasants. One of the first stories that I heard on arrival in Rumipata concerned an adviser who was anxious to begin plowing because the last date for planting was rapidly arriving. Rain, though, had lasted for a long time and the soil was not dry enough to allow a water-soaked hardpan to break up. Against the advice of a group of peasants assembled one day, the technician gave the order to begin plowing. All that the tractor succeeded in doing was to further turn the water-soaked soil into a gluey mass that baked hard as soon as the sun came out.

More importantly, some practices did result in considerable financial loss to the embryonic cooperative. One of the most crucial in retrospect was the choice of the *papa blanca* potato instead of the traditional *papa compis*. While yields from *papa blanca* are higher than the *papa compis*, it is used mostly in frying potatoes; when boiled in the usual way, it loses flavor.

Papa compis is always preferred by subsistence peasants for this reason over *papa blanca.* The problem arose when with profit in mind the technicians set a price of four and a half soles per pound of potatoes; this price was much higher than the market would bear, and as a result many potatoes went unsold and eventually had to be dumped.

Peasants were particularly concerned about potato production in the new cooperative because they believed that the provisional administration committee had assumed an enormous debt to pay off the assessed value of the haciendas that were expropriated and now administered by the cooperative. They were aware of the expenses involved in the widespread use of insecticides, fertilizer, and other inputs in the cooperative, the uncertainties of agriculture at eleven thousand feet and the importance of adjusting prices to meet optimum profit taking on the market. In their own household economies, they regularly held surpluses as long as possible to maximize profit for those surpluses they were able to market; thus, from their perspective, the inability of the technicians to understand price and demand factors of such a staple as potatoes was inexcusable. Peasants misunderstood, though, the strategy of selecting *papa blanca* over *papa compis;* the cooperative wanted to market the potatoes in the urban centers where *papa blanca* was preferred and *papa compis* was a rather unknown strain.

The plowing up of the hacienda lands and their cultivation appeared to be a rational long-term decision by PAC in bringing land into production that had been less than intensively used by the hacendados. It caused problems with the local communities, though, especially for those peasants who had been *yerbajeros* on the hacienda. The decision to cultivate potatoes meant that pasture land would no longer be available. As a result, the *yerbajeros* and others who had joined the cooperative to gain access to the fertile pampa pasture land were out of luck. Several members of the cooperative did send a formal request for land to be set aside but it wound up in bureaucratic channels and was not acted upon for months. In the meantime, when the dry season came and pasture became scarce, some cooperative members were forced to take their animals to the puna for pasture and effectively gave up their membership in the cooperative.

When land was set aside by the cooperative in recognition of the member's plight, further problems emerged. The land provided was to be used by members from not only Rumipata but from several other communities. This caused considerable dismay to the Rumipatiños; not only were relations strained between several of the communities, but the land was unprotected and open to the threat of cattle thieves who operated between the Pampa and the Apurímac River basin. Their solution to the problem was to build a hut on the land and take turns guarding the cattle during the night when the threat was greatest. Permission was required from cooperative officials for such action, and here the members were again frustrated. To get action on the land, they had to wait until the next general assembly that was some weeks in the offing. Meanwhile, several cows were stolen from the land. When a decision was finally made, it was poorly conceived from Rumipata's perspective. A man from another community was hired to guard the cattle, but suspicions were high that he at the very least turned his eyes away from the activities of the cattle thieves. Finally, one of the most respected members of the local committee, one of the last *personeros,* had his prize dairy cow stolen, capping a period of considerable ill will over the pasture issue.

Antapampa

During the latter part of 1970 and the first months of 1971, there were a considerable number of rumors about the eventual implementation of the agrarian reform. Expropriation proceedings had already begun on the adjoining haciendas, and it was generally accepted in Antapampa that expropriation was imminent. Expropriation of the major adjoining hacienda was nevertheless late in coming; a supreme decree was issued to that effect in the middle of May 1971, toward the end of the first wave of expropriations. While PAC did not seed a crop during the 1970–1971 growing season, there was barley growing there from the previous year. This was a result of the use of a mechanical harvester during the previous barley harvest that allowed enough grain to slide through to seed a new crop the following year. Much of the hacienda land, though, had been seeded in maize and provided forage for the cattle of the hacienda and *yerbajeros* after the maize was cut. The first effect then of the imminent expropriation was the lack of the maize pasture for cattle.

Most of the peasants that eventually joined the cooperative had been *feudatarios* on the hacienda; unlike the haciendas adjoining Rumipata and Tukiwasi, the failure to cultivate the land during the 1970–1971 growing season meant that there was no wage labor that many had thought would become available once the cooperative took over the land. Those peasants who joined the cooperative to obtain work were disappointed that no work was forthcoming. It reflected a lack of foresight in not cultivating the hacienda as in the other properties expropriated by the agrarian reform. Later, members began to express discontent publicly and some made inquiries to leave the cooperative and regain the funds expending in joining. Ex-*feudatarios,* though, who held usufruct plots on the hacienda continued to cultivate them in the uncertain situation, even if they seeded less because of a fear of having to give them up to the cooperative.

While operations had indeed been suspended on the local cooperative lands, when opportunities arose for work in other parts of the Pampa on cooperative land, obstacles emerged. Work was available in the northern end of the Pampa. However, Antapampeños did not like to work among strangers, and the difficulties of transportation and sustenance were great. Wages were often illusive and slow in coming. During one peak labor period when there was a call for Antapampeños to work on a potato harvest, they were promised a share in the profits of the particular harvest when it was sold, while members who resided in communities adjacent to the land were paid at the end of the week.

Tukiwasi

The Hacienda Tukiwasi consists of a series of sections as a result of the fragmentation described in Chapter 2. Only one section that I will refer to as Section A was expropriated in the initial stages of the agrarian reform; the remaining two sections still remained in bureaucratic channels even as I left the community in late 1972. During the 1970–1971 growing season, Section A was cultivated by PAC representatives in conjunction with the ex-hacienda administrator. The harvest was relatively successful. *Papa blanca* potatoes were seeded and considerable inputs in insecticide and fertilizer were applied; although the land had been cultivated by the owner prior to its transfer to PAC, informants reported that the harvest was good. A serious problem occurred, though, upon harvest when the potatoes were piled up without having been first graded into first, second, and third grades according to size. No buyers would bid on the potatoes in this form, and before the potatoes could be graded they rotted. The choice of *papa blanca* in this regard was particularly detrimental inasmuch as it has poor storing properties. As a result, a large amount of the harvest had to be given away as pig feed. These errors were regarded as serious by peasants whose knowledge of potato cultivation had been qualified by past experiences.

Summary

The activities of PAC and the cooperative at the local level was an important base against which peasants could evaluate the agrarian reform. The crucial activities described in this chapter were the economic activities involving exploitation of haciendas adjoining peasant communities. They provided a wealth of information concerning the motives peasants had for joining the cooperative and the overall trajectory of the cooperative as an economic unit. Through direct observation of these activities, peasants came away with a subjective evaluation of the risk involved in participating in the cooperative.

Unfortunately, their observations indicated a high degree of risk, caused by a number of factors. To some peasants, it was not altogether clear that their goals in joining the cooperative would be met. Pasture decreased under the spread of potato cultivation, and when pasture was available, the risk of leaving one's cattle on cooperative lands increased because of the threats of theft. The possibilities for wage labor never seemed to materialize: They were sporadic, located in land far from the community, and often seemed to be minimized by cooperative decisions to invest in labor-saving technology. To those peasants concerned with the trajectory of the cooperative, there was evidence that it would have a tough fiscal fight to maintain solvency. On the one hand, production and distribution decisions seemed to be made with a lack of understanding of key factors. Perhaps the best example was the widespread knowledge that much of the 1971 harvest had to be dumped because of the failure to select potato grades and setting too high a price to sell the potatoes. Certainly peasants misunderstood the questions of the agricultural advisers; while they were designed to uncover the local range of agricultural practices, they indicated instead to the peasants that the advisers had no more than "book learning" and even that was in short supply. Major decisions were made by "urban" mestizos acting in their capacity as advisers but having a major input into production and distribution decisions both short and long term in range. This bred considerable suspicion among peasants who had been exploited in the past by mestizos coming to them from the larger society. These were different faces than those of the mestizos who lived in the pampa; although they were aware of the structural changes Peru was undergoing, they had no reason to trust the motives of advisers from the larger society.

7

Peasant Decisionmaking: The Sociopolitical Context

While peasant beneficiaries were able to observe the short-term economic performance of the cooperative on the adjoining ex-haciendas, it was much more difficult to evaluate the organization of the cooperative as a model of production. The formal communications program was largely ineffectual; in its place informal brokers emerged at the local level to articulate the cooperative to the beneficiaries. Their success in this regard was qualified by the extent to which they reflected the collective interests of peasant beneficiaries. Covered in this chapter will be these and other themes that bear on the reaction of peasant beneficiaries on the local level to the sociopolitical aspects of the cooperative. An analysis of events in each of the three communities, during the period of initial contact with the then incipient cooperative, and, later, during its first year of operation, will be included.

Long-Term Economic Policy in the Cooperative

Long-term productive decisions are based on the maximization of productivity of the cooperative as a unit with the expectation that rewards will be shared among all of the members. It necessarily is complex, involving a hierarchical division of labor, a sophisticated, codified set of rules and regulations regarding its operation, and the delegation of responsibility. Most importantly, long-term production decisions are made through delegates elected from among nuclei of cooperative members. Local members, thus, neither participate directly in long-term production decisions, unless they are elected a delegate, nor in short-term production decisions, which are the responsibility of an individual or individuals delegated authority by the upper administrative level of the cooperative. In effect, the cooperative approximates the model of the agro-business in which production strategies are oriented to the maximization of profit of the entire unit rather than the survival of separate production units.

The contrast between the organization of long-term production and between the cooperative and the peasant community should be apparent. Long-term production decisions are made either in the assembly in independent communities or by the *mayordomo* in hacienda communities. In independent communities, they emerge through the tedious process of group decisionmaking. This process is based on normative rules that include face-to-face discussions of issues, and agreement is arrived at through consensus. Anybody who has something to say about an issue has the right and is expected to speak out before the others and present his case. Once a decision is reached through consensus, sanctions, both formal and informal, act to ensure that all peasants abide by it. Long-term decisions are oriented to the viability of the individual household over time: They are concerned with reciprocity, redistribution, and renewability of resources. Long-term policy does not normally involve the collective exploitation of resources it holds in common; rather it is concerned with allocating land to households on an individual basis. As Nicholas Georgescu-Roegen points out (1969:73), redistribution in peasant societies is done in order to equalize the opportunity of achieving an adequate household subsistence rather than equalizing income. While there is a sort of social welfare in Andean communities, close examination reveals that it is concerned with the disadvantaged: orphans, widows, the elderly, and the like. Georgescu-Roegen's comment regarding redistribution could include other functions of the supra-household: reciprocity, renewability of resources, and so forth. Their rationale is to maintain the "state of well-being" of the individual household.

The incongruity between the goals and formulation of long-term policy in the cooperative and the peasant community led to considerable confusion and misperception. The delegate assembly was perceived, especially by peasants from independent communities, as a rather large assembly that operated under the same rules which peasants were accustomed to locally. An old and respected cooperative member from Rumipata said that he had attended a delegate assembly meeting and raised his hand to express an opinion on an issue that was then being debated. He was interrupted and told that he could not participate in the meeting since he was not a delegate. He expressed a sense of frustration mixed with perplexity. "After all," he said, "who was making the decisions?" His frustration was based on the lack of understanding of the delegation of power to individuals to act on his behalf in the delegate assembly. The lack of comprehension of the delegate assembly extends to the function of the local committee. Ideally, the local committee acts as a sounding board in which members would instruct their representatives concerning their views on issues to be voted upon in the delegate assembly. The confusion over its role has seriously hampered its effectiveness.

Even if there had been some understanding of its role, the composition of the local committees constrained its decision-making abilities. Because of its composition, many local committees allow the younger, literate, and more aggressive members to guide the course of discussion on topics and issues that often are rather sophisticated in nature. The older, more conservative members are at a disadvantage in this regard. The situation is different in the assemblies where only household heads may participate and the resulting gerontocratic bias is correlated with a status-prestige hierarchy based on the civil and religious *cargo* sponsorship. These two aspects of the dynamics of the local communities have obstructed participation. Older cooperative members were ill at ease in the meetings of the local committees.

As discussed in Chapter 5, some effort was made to establish intermediary linkages at the regional level through "peasant leaders." A similar process occurred at the local level. In some instances, mestizos who had held key roles in the pre-reform social structure of the local community were allowed to continue in these roles. In other cases, different factors were instrumental in the emergence of an individual or set of individuals. In general, though, in each community there was a process of selection of brokers who were instrumental in the manner in which the cooperative was perceived by the beneficiaries. As in the selection of brokers at the regional level, political competition occurred at the local level. In effect, the structural changes brought about by the agrarian reform and associated legislation caused new alignments in the relative position of interest groups at the local level and important changes resulted in the makeup and behavior of the intermediary groups.

One group particularly affected by the shift in power was the *pequeños propietarios,* or mestizo middle holders who had bought or otherwise acquired a sizable quantity of irrigated land but who did not participate in the affairs of the community other than to recruit labor. Their traditional basis of support was the coalescence of their interests with members of the regional mestizo stratum: other middle holders, town dwellers, hacienda administrators, and the like. In turn, they depended on juridic-political links with the hacendado elite in Cuzco, maintained through regular social interaction with its regional representatives located in Izcuchaca, the mestizo trade and communication center of the pampa. With the agrarian reform, and the elimination of mestizo control over land, the base of power of the middle holders was effectively removed.

The economic livelihood of the middle holders depended on secure access to land and the recruitment of labor on favorable terms from residents of the independent peasant communities. Although the quantities of land they held were usually below the minimum legally required for expropriation there was still considerable anxiety that it would be expropriated. Even if their land did fall below the minimum, Article 126 of the Agrarian Reform Law permits land to be expropriated from neighboring holdings in and adjoining peasant communities and the minimum expropriable size reduced when peasant communities lack sufficient land to cover their needs. This may occur even though the region has not been declared an agrarian reform zone. Also properties that have had a history of "social problems" with adjoining communities are subject to expropriation.

The result was a shift to residence and participation on the local affairs of the peasant communities. Many new two-story homes, bright with paint and tin roofs, began to sprout up in the Pampa de Anta marking the properties of middle holders who had built the homes to establish their residence in the community. Middle holders began to attend assemblies of peasant communities, speak out on issues, offer to assist the communities in projects, and generally to be active in the life of each community. Ultimately, as occurred in January 1972 in Rumipata, they petitioned for status as *comuneros* and were voted into full membership.

Another group was made up of individuals who had held

land in their communities of origin but had migrated either temporarily or permanently to other communities or urban centers. Their land was threatened by clauses in the agrarian reform law prohibiting indirect forms of tenure such as sharecropping arrangements commonly used by migrants to retain control over their properties during their residence outside the community. Many migrants did not even maintain indirect arrangements for cultivating their properties and left them to fallow during their absence. These clauses of the agrarian reform law were designed to end these practices that were obviously inequitable in relation to the landless segment of each community.

As a response to these concerns, there was a rapid and increased rate of return migration to communities of origin following 1969. An insight into the magnitude of return migration can be see in Figure 7–1. Approximately 14 percent of the total sample interviewed indicated they had returned to the community from residence in another locality as a result of the agrarian reform. In most cases, they were migrants from a large urban center, usually Lima.

Figure 7–1. Return Migration Following the 1969 Agrarian Reform

	Total Interviewed	Return Migrants	Percent
Rumipata	46	8	17.4
Antapampa	29	2	6.9
Tukiwasi	13	2	15.4
Total	88	12	13.6

The decision to return to their communities of origin, to reestablish residence, cultivate their plots, and to participate in community affairs was a portentous decision on the part of the migrants. For many, especially those who had returned from residence in Lima, it was a quite difficult decision implying that one was dissatisfied with city life and must face the recognition of downward mobility implied in a return to the countryside.

The impact of a rapid and increased rate of return migration combined with a shift in the orientation of mestizo middle holders varied considerably in each community. In Tukiwasi, there were no middle holders except for MCS whose position in the community was firmly entrenched. The rate of return was much lower in Tukiwasi than in the other independent communities; the rate and subsequent impact of return was greater on the other hand in Antapampa and Rumipata.

In Rumipata, the presence of a significant segment of return migrants and several middle holder families living in one sector of the community was an addition to another immigrant segment of the population dating from the last two decades. Part of the latter stream can be attributed to Rumipata's position with respect to the Pampa de Anta, the high puna ecological zone, and other population centers to the south of the Pampa. Until a road was built to Cotabambas in the Apurímac River basin in 1962–1963, Rumipata functioned as a port of trade for the exchange of products from the communities in the high puna zone between the Pampa and the Apurímac basin and the more temperate maize zone of the pampa. Wool, dairy products, *chuño,* and other puna products were brought to Rumipata to be sold or traded to shopkeepers and/or middlemen who came from Cuzco. During this period, some of the entrepreneurs attracted to Rumipata eventually bought land and settled or married into the population. Other migrants from communities as far away as Cotabambas came to the community before the road was constructed to be closer to the Cuzco highway and to have access to the school in the community.

The presence in Rumipata of a rather substantial "nation-oriented" group placed a further strain on the traditional sociopolitical integrative mechanism. As a response to the problems of identification of the newly created group and the inability of the mechanisms of the community, including envy, threats of witchcraft, rumor and gossip, to control the process, a voluntary association, the Club "San Marcos," was organized in 1970 by some recently returned immigrants from Lima and soon began to recruit other "nation-oriented" individuals into the association. Within the next two years the Club became one of the most active associations in the community, organizing a soccer team, conducting fund-raising activities, holding regular meetings, and pooling contributions of members to provide funeral benefits for deceased members. One of the most important functions of the Club has been in initiating a special fiesta based on an amalgam of coastal, highland mestizo, and indigenous cultural elements (Guillet and Whiteford 1974). The ritual involves sponsorship by male and female members on a rotating basis, thus re-creating in a transformed guise the features of the *cargo* systems traditionally associated with indigenous sociopolitical organization.

While the formation of the voluntary association as a vehicle for the interests of certain of the nation-oriented groups in the community did occur, it was insufficient to contain the centrifugal disequilibrating forces unleashed by the agrarian reform. Informants from Tukiwasi and Rumipata, which are adjacent, and tied into various interactive spheres, were constantly contrasting the situation in "united" (*unido*) Tukiwasi with Rumipata.

The situation was not quite as bad in Antapampa, where an earlier division of the community into an upper Antapampa and a lower Antapampa had quelled some of the disequilibrating tendencies building up during the last few decades. Antapampa was particularly susceptible to the influence of a middle holder group larger there than in any of the other two communities. Since many of the middle holders had properties located within Antapampa's boundaries, they were particularly concerned with maintaining good relations with the residents. In the past, they had never bothered to participate in community affairs and had never tendered membership in the community as something of value.

I will now turn to a short review of the events in each community at two points in time: (1) the period from the initial contact with the community by representatives of the agrarian reform and, (2) the first year of operation of the cooperative. I will look particularly at the selection and performance of the broker and the relation between the performance of the broker and the internal dynamics of the community. The nature of the group within which the actor is located is important to an understanding of behavior under uncertainty. In the peasant context, the local community is the immediate reference group; social changes that either integrate or disintegrate the community would ultimately affect the ability of the community to influence the individual's perception of the outcome of some contemplated action, in this case the decision to join, or participate in, the cooperative.

The relation between the broker and the leader that has

been alluded to earlier is extremely important in this context. Inasmuch as the broker is structurally situated in the community to assume the role as a leader, and the closer the correlation between broker and leader, then the greater the flow of structurally relevant information. Leadership is particularly important in conditions of uncertainty especially where there is a need to make decisions that are also innovations (F. G. Bailey 1970:58).

Rumipata

The initial contact between agrarian reform representatives and the community occurred when the community was requested to elect delegates to attend the seminar in the summer of 1970. Two delegates were elected, attended the seminar, and returned to the community to report their experiences. There was a general satisfaction with the plan as outlined by the delegates, and from all reports many peasants wanted to join the cooperative. As was mentioned previously, the plan involved parcellation of expropriated land in parcels of three hectares each to cooperative members. The cooperative would then assist in providing labor for members to use and function as a service cooperative, dispensing fertilizer, insecticides, and other capital inputs at favorable prices. When this plan was rejected by the Ministry of Agriculture, there was considerable disillusion in the community; it was reflected in increased cynicism directed toward the government, and accusations were made that the original community delegates lied to them.

Informal networks of information added to the confusion. A number of rumors were spread by mestizos, in particular, middle holders who felt threatened, that the agrarian reform was designed to exploit the peasants. They said that the census then being taken to determine which peasants had a legal right to be declared beneficiaries was in effect a plan to confiscate all of the livestock owned by peasants. This was the rationale, according to the rumor, behind the seemingly unending, detailed questions concerning the number, type, sex, and age of livestock owned by the peasant household. In almost every reference to the imminent agrarian reform, mestizos attempted to persuade peasants that it was another ill-founded and probably exploitative scheme devised by the government to use the peasants. In many instances, having no other viable source of information, peasants were ill at ease by the rumors spread by mestizos.

During the initial contacts, agrarian reform representatives used the communal assembly to explain their program to the assembled peasants. Such a forum had regularly been used for public discussion concerning events of interest to the community. Sensing the need for a more propitious link to the community, through its informal network of communication, the representatives were especially dependent upon AC. AC was a schoolteacher educated in Cuzco and Puno who had been born in the community but for the last several years had lived outside the community. At the time of the agrarian reform, he had been working for the Ministry of Education in Izcuchaca but maintained a family and a residence in Rumipata and made regular trips there. When elections were held under the new peasant communities law, he was elected president of the vigilance council. Another recently returned migrant from Lima who associated with AC was elected president of the board of directors. These two individuals were basically *mistis* of the newer variety; the rest of the political officers were held by members of the *runa* stratum.

These three individuals were elected because they displayed an ability to deal successfully with institutions of the larger society. They were literate, fluent in Spanish, and had lived for several years in urban centers. Ordinarily holding public office is not looked upon with much appeal. Considerable time must be devoted to community affairs, and there was always the possibility that one would be called down for a real or imagined dereliction in office. For these three individuals, there were rewards to be had from holding office. As return migrants, they had to become established in community life and to legitimize their presence to such an extent that they could defend themselves against the threat of expropriation contained in the agrarian reform law. Election to office, if only in name, was one sure way to ensure that there was no argument as to their rightful participation in community life.

The schoolteacher soon began to assume the major burden of office: he set dates for and ran the assemblies, called faena, labor parties, and supervised them, laid out plans for a new market, and represented the community in negotiations with a hacienda located within the communal boundaries. And he directed the annual redistribution of communal land.

Using his position as an officer in the political organization of the community, he expanded his niche to include a role as spokesman for the agrarian reform: internally to the community and externally to the bureaucrats charged with administering the reform. Although he was barred from actual membership in the cooperative by a clause prohibiting civil servants from membership, his wife joined and was elected delegate to the general assembly from Rumipata. As membership was taken during May of 1971, a core group of members began to gather around the personage of the schoolteacher. This core group is quite important for the perception of the cooperative by the rest of the community.

The core group was made up to a large extent of friends, kinsmen, and members of the same residential sector in which the schoolteacher lived. There were 22 co-op members in the group that joined out of a total of 265 household heads that were eligible under the census of the preceding year. To a large degree, cooperative members were wealthier than the majority of peasants in Rumipata. In a sample of those cooperative members for which data from the 1970 census is available (17 out of 22), there was an average of .53 hectares per household of irrigated land in comparison with .46 hectares per household of irrigated land for the community as a whole, and 2.73 hectares per household of puna land as compared with 2.15 for the community. There was a lack of participation in the cooperative of the peasants who theoretically would most benefit from membership: the landless and the land hungry.

The rather disappointing participation of peasants in the cooperative can be explained by the fact that most peasants adopted a wait-and-see attitude: (1) to determine the subsequent developments and success of the cooperative, and (2) to watch the manner in which the local membership, notably AC and members of his faction, responded to the responsibilities of cooperative membership.

The First Year

During the first year of operation, the local committee in Rumipata was inextricably tied to the fate of AC. Although he was formally barred from participation as a civil servant, he was active in both the local committee, where his wife was elected delegate, and in the central offices of the cooperative for a short time as secretary to the first elected president of the cooperative. The latter position as well as other contacts he cultivated enabled him to become a spokesman for peasant interests. When President Velasco visited the pampa in October 1971, AC was in an elite group of cooperative officials who were introduced to the president and received special recognition. Such a highly visible presence resulted in reprisals brought against AC by the Ministry of Education, his employers at the time. Political activity was prohibited by the ministry policy, and although many schoolteachers do engage in local politics, AC's open courtship of the reform bureaucracy was a public breach of policy that had to be contained. As a result, he was transferred to teach in a small, remote peasant community some hours ride by horse from the pampa. Nevertheless, he returned to Rumipata on weekends and other occasions to carry out his business and maintain a presence in the community and in the cooperative.

In the last few months of 1971, AC began to have problems as an elected official in the political organization of the community. As previously mentioned, while his formal position was that of the president of the vigilance council, he in effect assumed responsibility for all of the internal community activities and served as community representative externally in contacts with the Peasant Communities Agency, visiting Cuzco approximately twice a month on community business and Lima on one occasion in 1972.[1]

AC's position began to deteriorate during the latter part of 1971. First, he was criticized by the older members of the community when he distributed land without regard to the rules concerning the rotation cycle. This was a serious charge that implied his lack of knowledge or concern for the renewability of the highly critical puna lands. He then attempted to display his ability to negotiate with *pequeños propietarios* in

1. The president of the administrative council never took an active role in community affairs and in December 1971 notified the assembly that he had to return to Lima because his wife was ill. There was an underlying suspicion that he had always planned to leave, especially since his family had continued to reside in Lima and he had made only trips over the years to Conchacalla for short periods of time, never really establishing a household there. It had little effect on local politics, inasmuch as he was replaced by the vice-president who raised no objection to AC's control of the assembly.

behalf of the community. Although he did succeed in this regard, he eventually opened himself up for more criticism. There were two properties located in the community involved in these negotiations. The first was a small holding that AC felt was subject to expropriation, since the owner lived in Cuzco and administered his property through a *mayordomo*. AC made a formal request signed by the members of the community to the Ministry of Agriculture that the property be expropriated and turned over to Rumipata. The community, in turn, would operate it as a chicken enterprise with *comuneros* supplying labor. To this end, AC requested a loan of thirty thousand soles and two scholarships to train *comuneros* to manage the enterprise. This plan was brought up continuously by AC for discussion at assemblies, but it was never acted upon during the year.

The second piece of property was owned by another middle holder who had attempted to sell it to a priest in Anta for a ridiculously low sum, sensing that it would probably be expropriated soon. When the plan was publicly revealed, the community protested, and through a series of negotiations led by AC, the property was finally sold to the community. AC's initial plan, again in agreement with community members, was to farm the land using faena labor of the community. A fund for capital inputs would be created by donations of ten soles and two kilos of potato seed from each household. This plan never materialized. What did ensue is not clear; when I visited the lands they were being farmed in sharecropping arrangements with the community as owner and four of the wealthiest individuals in the community, one of whom was AC, as tenants. AC was farming his plot using free labor recruited from peasants who had missed a faena and therefore owed work to the community. There was considerable criticism of AC for his use of community labor for personal ends.

Finally, a major conflict between AC and the schoolteachers in Rumipata brought all of the underlying tensions to a head. During the latter part of 1971, a complaint was made that schoolteachers were lazy, drinking excessively, and playing soccer instead of teaching. In retaliation, the schoolteachers called a meeting of the fathers of the schoolchildren and arrived at a consensus that AC's activities as a civil servant and a political officeholder were noncompatible. A formal charge against him eventually reached the Ministry of Education in Cuzco, resulting in a running battle of community politics without any immediate resolutions. It succeeded in bringing to the front of public debate, both in assemblies and in faenas, the split between two factions that had emerged: supporters of AC, and a faction opposed to AC made up of schoolteachers and a group of return migrants and *pequeños propietarios* who had recently entered into local politics. By early 1972, the dispute had reached such an innuendo of threats, counterthreats, and, in some instances, accusations of witchcraft and envy, that it overwhelmed community life.

The annual vote (held in January) on peasants petitioning for community membership is especially illustrative of the maneuvering process that was then occurring. This vote is the last stage by which the new residents of the community become officially recognized as members. In the first step of the process, individuals undergo a period of from five months to a year of "observation" during which they are supposed to indicate their interest and worthiness by attending community functions and participating in the day-to-day life of the community.

During the meeting, two groups presented themselves to be voted upon. The first group consisted of those who had been under observation for the given period—mostly peasants, some returned migrants, and a few mestizo middle holders. Each individual stepped forward before the assembled *comuneros* and made a short plea to be granted membership. AC presided over the process, commenting on each individual and clarifying the law as set forth in the statutes governing peasant communities. He then asked in Quechua those who voted in favor of the postulant to raise their hands. Approximately half to two-thirds of the peasants present raised their hands each time; this was taken by AC to be a majority. No count was made of those opposed, and only once was there a discussion about the merits of a particular individual. In the latter instance, there was some discussion about an individual who had never attended assemblies and faenas; nevertheless, he was eventually voted into the community. When one of the leaders of the faction opposing AC came up to be voted upon, AC first commented negatively on the individual saying that by law he should not have *comunero* status since he had land in another community; he then changed the pattern by asking first for a vote of those who were opposed to the individual to raise their hand. He was not able to get a majority, and the man was voted into membership.

The second group was mainly middle holders who had not passed through an observation period. Nevertheless, they were voted into membership. In one case, an individual who had never even appeared before in the community was voted upon favorably.

The factioning in the community and criticism made of AC's performance by middle holders and return migrants considerably weakened his ability to represent the local committee of the cooperative with the central offices. It became eminently clear that AC was attempting to use his position in the community and in the cooperative for personal aggrandizement and that his interest in the well-being of peasants and cooperative members was secondary. AC had to expend considerable energy to protect himself and his supporters against criticism by other individuals competing for these crucial niches in community life. As the internecine struggle gained momentum, AC began to lose hold, factioning increased, and it was obvious that AC could not deliver on the promises that had brought his support initially.

AC's demise was reflected by a lack of interest in the cooperative both by members and nonmembers. Only three to five members worked with any regularity for the cooperative while the rest ignored periodic calls to work. This is somewhat understandable given the composition of the original group of members. There was a high proportion of members who joined out of support for AC but who held primary occupations outside of agriculture that brought in a stable and sufficient income. Their interests were not compatible with those of the few members who had joined because of perceived opportunities for wage labor and capital investment in the cooperative. The frequency of meetings declined from three or four times a month to sporadic meetings held only when the occasion demanded. The delegate from Rumipata, AC's wife, stopped attending meetings of the delegate assembly.

Antapampa

Initial contact between the representatives of the Ministry of Agriculture and the community began during the summer of 1970. They approached the peasants in a regular assembly format to explain the cooperative that was then being formed. They gained entrance to the assembly by contacting the head (RP) of one of the older and most prestigious families of the community. RP was known to functionaries in Izcuchaca and Cuzco, as well as in his community of residence. He was asked by the representatives to help them explain the cooperative concept to the community. He agreed and during the meeting acted as the go-between.

The choice of RP was certainly fortunate for the Ministry of Agriculture. His family was by far the most important, the wealthiest, and the most prestigious in the community. RP had been president and treasurer of the community, had held all of the major *cargos,* and in addition had represented the community in all of the commissions sent to provincial and departmental offices. Antapampeños trusted RP as an individual who had lived in the community most of his life and who held important positions of duties in a responsible manner.

Considering the uncertainties, though, there was a respectable response in May to the recruitment of members for the newly organized cooperative. Based mostly on the support manifested by RP, thirty-seven members joined the cooperative and paid one hundred of the five hundred soles required for admission. Among these members, there was a considerable bias toward *comunero/feudatarios;* over half (54 percent) were in this category. As in Tukiwasi, this was an indication that both irrigated and pasture land belonging to the hacienda was now owned by the cooperative, and that *feudatarios* would have to join the cooperative in order to continue to have access in one form or another of their plots. Not all *comunero/feudatarios* joined though; there were a considerable number that did not.

The First Year

The local committee in Antapampa met only sporadically during the first year, mainly to discuss issues communicated to them from the central offices of the cooperative: for example, meetings were held to talk about the call for labor to meet the potato harvest, to organize a contingent of Antapampeños to attend the fiesta of the cooperative, and to contribute to a fund-raising event, a *kermes,* sponsored by the cooperative.

Only a fraction of the members worked regularly for the cooperative.

RP's role as a spokesman for the cooperative was eclipsed during the year due to a series of factors. He had accepted the role initially out of a sense of obligation to the community and as an aspect of his own position of prestige in the community. Following the initial contact between the community and representatives of the agrarian reform that led to the enrollment of Antapampeños in the cooperative, RP retreated from actual contact with the cooperative bureaucracy and maintained his positions vis-à-vis the cooperative only as required by the community. Other members elected to the positions in the local committee took a somewhat more active role in associating with middle and upper level officials, but their motivations in doing so coincided with RP's reasons for not doing so. They were concerned with actually working for the cooperative in order to receive the wages that were paid and in the process gain prestige for exhibiting an ability to interact with other upwardly mobile peasants. RP, on the other hand, had already gained prestige through traditional mechanisms within the community. To him neither prestige nor the opportunity of working for the wages of a day laborer were sufficient motivations to maintain active contacts with the central office of the cooperative.

Other groups in the community viewed the opportunity of acting as a go-between for the community and the state agencies as an appealing position. Their continual criticism of the agrarian reform at all levels was designed to show them as more capable than the elected officials of the local committee. At the same time, another set of factors complicated the ability of officials of the local committee to articulate their needs. Most of the members elected to positions in the local committee also held important positions in the political organization of the community. The nonmember population of Antapampa sensed a conflict in the allegiances of these individuals. For that reason, they demanded that one position or the other be relinquished. As the annual elections in October arrived for positions in the vigilance and administrative councils, the entire community felt the need to consult the Peasant Communities Agency in Cuzco to get advice regarding an individual's eligibility to hold membership in the cooperative and a post in the community concurrently. The confusion was a result of the perception in the community of two quasi-corporate groups: one made up of the cooperative members, referred to as *socios,* and the other made up of noncooperative members referred to as *comuneros.* During community meetings when both groups were present, the *socios* were placed on the defensive because of the criticisms levied by mestizo middle holders and other individuals who felt qualified to comment on the manner in which the agrarian reform was affecting the community. The lack of a motivated spokesman who was qualified to make statements about cooperative policy was apparent; this lack was exploited by groups interested in attempting to place themselves in the position of spokesmen for the community. They began to point out the inadequacies of the cooperative and the inability of the local committee to perform.

An example of this inadequacy is found in an assembly meeting that took place in the early summer of 1972. During the assembly, it was suggested that a plot of land should be appropriated for a soccer field for students. A cooperative member said that a plot of land on the local ex-hacienda could be requested from the cooperative. The discussion then turned to whether the community should present a formal request (*solicitud*) to the cooperative asking for the land. Some *comuneros* said this would not be accomplished since the request would end up within the cooperative bureaucracy. Two factions began to emerge: the first, headed by an official of the local committee of the cooperative was on the defensive. He argued that the local committee should handle the request and that the cooperative would grant them and no other group the land. The second faction consisted of a group of mestizo middle holders who said that they would talk to a middle holder who owned another piece of land in the community that they felt would be more appropriate. The cooperative, they said, would never give the land to the community. They further implied that only they could get the land from the middle holder. These two factions dominated the discussion while the majority of the *comuneros* watched on the sidelines. There were clear status differences that could be discerned in the discussion: the officer of the local committee of the cooperative used *señor,* a term of respect, in addressing the leader of the middle holder faction, while the latter used *tú*, in this context a demeaning form, with the cooperative member.

The result of the factioning was to present a dilemma to the Antapampeños. As a relatively poor community, they were interested in the cooperative as a stable and secure source of work at relatively high wages, as a viable economic alternative where there has traditionally been a lack of alternatives. Yet, their experience and the performance of the cooperative left much confusion in the minds of the peasants. If there had been trusted and motivated individuals linking the community and the cooperative who could have withstood the attacks of mestizo middle holders, then there would have been a more secure base upon which to estimate the rewards to be had from cooperative membership. Such a linkage did not emerge though, and there was general uncertainty concerning the future course of the cooperative.

Tukiwasi

In order to understand the local level response to the agrarian reform in Tukiwasi, we must recall the manner in which the community was structured during the hacienda period. As will be seen, it produced a dilemma for the agrarian reform bureaucracy in their initial contact with the community.

One individual (MCS) has effectively wielded independent power within the local community in making decisions regarding virtually all aspects of community life. Although his major source of power stemmed from his position as representative of the hacienda, he was also *teniente gobernador*, keeper of the keys of the church, catechism instructor, and maintained fictive kinship ties of a paternalistic nature with virtually all of the households in the community. *Feudatarios* had had no occasion to question the authority of MCS and yielded to his direction in organizing activities in the community, seeking his opinion in understanding the events that were occurring in the Pampa de Anta and Cuzco. They viewed MCS as the only individual both capable and trustworthy enough to act in their behalf. While MCS's position can be traced to the role of the hacienda in the regional political economy, it was not perceived by *feudatarios* as exploitative but the best that they could obtain under the circumstances. MCS in his own activities had garnered their trust through long residence in the community and a clear interest in the well-being of the *feudatarios*, going beyond the minimum required of paternal behavior.

The problem presented to the agrarian reform bureaucracy was the manner in which MCS's role was to be recognized in their dealings with the community. On the one hand, they clearly perceived MCS's relation to the community and his traditional function in channeling information to the *feudatarios* concerning events originating in the larger society. It followed that MCS could play a crucial role in interpreting the agrarian reform to the *feudatarios*. On the other hand, MCS was a mestizo, a property owner, and a representative of the institution that the agrarian reform wanted to replace. The solution to the dilemma as acted out by both the representatives of the agrarian reform bureaucracy and MCS is a key to an understanding of the response of Tukiwasi peasants to the cooperative.

Sensing the imminent changes that were about to happen and the lack of an internal structure in the community to deal with the changes, MCS began to prepare the *feudatarios* for self-government. In August 1969, he called a meeting of the community in which he explained as best he could the agrarian reform then in its initial stages and the necessity for the election of local community representatives to the reform bureaucracy. This was the first time when the community was to act in an autonomous manner and in effect recognize itself as an independent decisionmaking body. Nevertheless, during the meeting and subsequent meetings, MCS called the meeting to order, carefully explained the topics that needed to be discussed, and organized formal procedures such as the election. Representatives were elected during this meeting.

The next formal meeting of the community was held in June 1970. MCS had been contacted during the previous month by the division of peasant communities to place before the community a set of options open to it: The first option was to form the community into a cooperative of services along the lines of the national cooperative law; the second was to file for recognition as an officially recognized peasant community. After a discussion led by MCS, all of the assembled *feudatarios* voted unanimously to become a recognized peasant community. Two representatives, *gestores*, were elected. Papers were drawn up and the formal process of recognition was begun.

As the summer of 1970 continued, MCS continued to play a seminal role in the negotiations between the community and the reform bureaucracy. In the understanding of the *feudatarios,* MCS had been "contracted" by the cooperative bureaucracy to serve as administrator of the Tukiwasi land that was in the process of being expropriated and distributed to the cooperative.

They felt that he would promote their interests in these negotiations. As it turned out, MCS was able to arrive at an arrangement to the mutual satisfaction of the community and the agrarian reform bureaucracy. This arrangement was a strategic one for the community. *Feudatarios* who elected to join the cooperative would continue to work on the lands of the Tukiwasi haciendas as well as any other lands that belonged to the cooperative. They would be paid the cooperative rate of twenty-four soles a day as any other member of the cooperative. But, unlike the general arrangement, no cooperative members from another community could work on the cooperative lands of Tukiwasi. The entire administration of the local enterprise would be in the hands of MCS and he would represent the community in the governing bodies of the cooperative.

In another area, MCS set up a format through which the Radio Forum program could be effectively disseminated in the community. In an assembly in August of 1970, the community was divided up into three groups of eight to nine members each with a "president"; these groups were to listen to the program on the days it was broadcast, and then interact within the group to discuss the issues, raise questions, and then seek answers from MCS. Since that time, the community has met regularly, usually on Friday mornings, to hear *en concert* the program.

Response to the call for membership from Tukiwasi was high; thirty-six household heads joined the cooperative. All had been *feudatarios* on the hacienda from before the reform. Not all *feudatarios* joined the cooperative. Those *feudatarios* who did not join continued to have contractual relations with a segment of the hacienda in the community not yet expropriated.

The only problem that continued to obscure the arrangement was the position of MCS. It became increasingly apparent that some accommodation to MCS's role in the arrangement would have to be made; his obvious mestizo status was clearly a thorn in the side of the agrarian reform bureaucrats who endeavored to create a post-agrarian reform social structure free from any hacendado influence. Because of this problem, a meeting was held in April 1971 between representatives of the Ministry of Agriculture and assembled *feudatarios*. Ostensibly the issue at hand was the complaint presented by the *feudatarios* that MCS was not able to join the cooperative because he was a middle holder. The solution came in another meeting the following month when MCS was unanimously voted into the community as a *feudatario*, "with a grand desire to join the cooperative." This accommodation was accepted by the Ministry of Agriculture because of the recognition that MCS was playing a crucial role in the reaction of Tukiwasinos to the cooperative then in formation.

The First Year

A review of the cooperative's operations in Tukiwasi during its first year reveals few major changes. While the community entered into a period of transition, the basic socioeconomic structure of the hacienda community continued to prevail.

The manner in which work was organized during the hacienda period has continued as the basis for recruitment and organization of work under the cooperative. Time is still allotted during a weekly meeting; the only difference is that a proportion of the total time is now allotted "to the cooperative" rather than to work on hacienda lands. Other time is still allotted to labor obligations on non-expropriated hacienda land and the remaining time is set aside for work on individual plots. MCS's role has not changed at all in this context even though he is no longer referred to as the *mayordomo* but as an official (*dirigente*) of the cooperative.

Work is not limited to only local hacienda properties; whenever labor is needed on other cooperative lands, a work order is sent to MCS who then will either call a meeting or wait until the next weekly meeting to set aside time for the work requirements. On the day when labor is requested, a truck arrives in the morning from the central offices of the cooperative to carry Tukiwasinos to their destination. They have gone on occasion as far as Chamancalla, at the other end of the Pampa, to supply labor needed for a large potato harvest.

Unlike the situation in Rumipata and Antapampa, Tukiwasinos do not object to being trucked a considerable distance to work for the cooperative. The difference in this case is that they go as a unit and often nonmembers, young girls in particular, will go along as contracted day labor. There is an apparent air of festivity; the security and solidarity of working with one's kinsmen and friends eases the stress of working in strange surroundings. This solidarity extends to other group activities carried out by Tukiwasinos. Their winning soccer team and experienced dance group is a source of prestige. As regards the cooperative they are quite aware that they are members and are happy with that status. Unlike the anxiety met within Antapampa and the hostility manifested in Rumipata, Tukiwasinos are pleased that they have adapted successfully to the cooperative.

The key to their success, which they readily admit, is MCS; he has successfully transformed the hacienda structure into a "cooperative" organization. Work orders issued by the cooperative are perceived as labor obligations due the cooperative in exchange for the individual plots they still farm. The fact that they are receiving a high wage for their labor is a token, in their view, of the successful work MCS has done for them in their behalf.

One manner in which the cooperative has been meshed with the pre-agrarian reform hacienda organization is in the meetings of the local committee of the cooperative that are coterminous with meetings of the community. Topics discussed in these meetings reflect activities of both the cooperative and the now autonomous community; it is difficult in this sense for a nonmember to escape the issues confronting the community and its role in the cooperative. Although individual Tukiwasinos do consciously make a distinction between a cooperative member and a noncooperative member, this distinction is a disruptive tendency within the community. To resolve this tension, strong social controls are forcing the minority of noncooperative members to reconsider their status and ultimately to join the cooperative. These controls surface on issues that are reflections of the new rights and obligations of the cooperative members. For instance, a common source of conflict during the field period was the fact that some nonmembers continued to pasture their livestock on the lands of the hacienda that now belonged to the cooperative. One informant said that members were envious of him for doing this and he had decided to join the cooperative to legitimize his pasture practices.

MCS's behavior is "altruistic" in relation to Tukiwasi. He has made a sincere effort to prepare the peasants for self-government and autonomous representation. An example of MCS's concern is his encouragement of the Radio Forum program to enable peasants to comprehend the issues of the agrarian reform. In assemblies, he continually stressed the importance of listening to the program, supplied a radio from his own home, and answered all of the questions during the sessions. When attendance declined during the early months of 1972, he brought the issues to the attention of the community and through his influence a fine was levied against nonattendance.

He has attempted to instill pride among the Tukiwasinos in being members of the cooperative. When the first anniversary of the cooperative in Anta was being planned in mid-1972, a series of dances was commissioned from each community in the Pampa. MCS encouraged the idea and about three weeks prior to the event, he and his daughter, the schoolteacher, held meetings at night with a group of dancers from Tukiwasi. They learned a mestizo dance taught by the schoolteacher that later won acclaim at the anniversary celebration.

In comparison with the other two communities, Tukiwasi strikes the observer as much more integrated. This feeling is shared by both Tukiwasinos and residents of other communities alike. Tukiwasi is "together," *unido,* they say, unlike Antapampa and Rumipata. This perception reflects the lack of structural change that would have upset the equilibrium between MCS and the community and his continuance in the broker role in the post-agrarian reform situation. MCS in this regard has been fortunate; he has had no competition from other individuals or groups, such as the middle holders or return migrants, found in other communities.

Tukiwasinos are well aware of MCS's special role in the community and the fact that there is no one among them capable of handling the responsibilities that MCS assumed. This issue will be perhaps the most important issue in the community's future: MCS was told in the summer of 1972 by the Peasant Communities Agency that he would have to leave the community because he was a middle holder with no business in local politics. This move was protested by community members and had caused much anxiety. They were well aware of what would ensue if MCS were to leave. He had no choice, however, and, as I left, MCS and his family were preparing to move to Lima.

The Three Cases Compared: Selection and Performance

In order to understand the process of brokerage, one should make a distinction between the selection (the informal process through which a broker emerges) and performance (his role in transmitting information to peasants at the local level).

Figure 7–2 illustrates the steps taken in the selection process.

Figure 7-2. Representation of Broker Selection Process

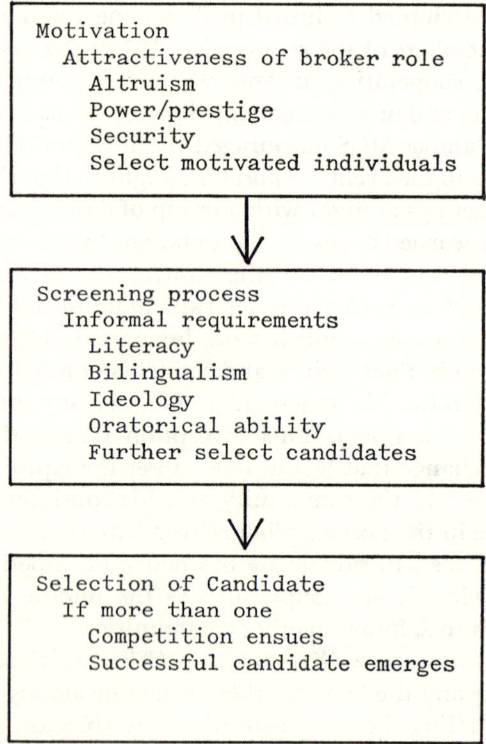

The first step involves motivation. We refer here to the complex of motives that, including altruism, economic security, and power and prestige, attract prospective spokesmen. The second step is an informal sorting of the qualifications of motivated individuals. Qualifications are those traits that are necessary to successfully act out the role. For example, to interact in the institutional setting of the agrarian reform bureaucracy one must be literate and fluent in Spanish. In the local community, on the other hand, literacy is unimportant and conversational ability in Quechua is a necessity. An ideological commitment to the goals of the agrarian reform and an oratorical talent are other requirements. The last step before a spokesman emerges is competition, only if there is more than one qualified candidate.

There is considerable variation in motivation among the three individuals we have described. AC sought his position as spokesman in order to legitimize his presence in Rumipata and thus safeguard his property against the possible threat of expropriation. Being spokesman also meant access to resources, such as free labor in the community and a lucrative salary on the payroll of the cooperative. In Tukiwasi, MCS similarly saw his role as a defense against threats to his privileges as a mestizo middle holder and *mayordomo*. However, he did display a genuine concern for the community and did make efforts to assist the ex-*feudatarios* in the difficult transition to self-government. In Antapampa, RP had long before achieved a position of respect and power and was under no threat from the agrarian reform. He had represented Antapampa in negotiations with government agencies on numerous occasions and saw the agrarian reform as another situation where his experience and talent would be needed for the good of the community.

The next step in the emergence of a spokesman is an informal screening process to assure that candidates meet qualifications set by the local level and the institutional setting of the agrarian reform. For example, the promoter is taught to approach the most influential peasant in each community in order to enlist his support in motivating beneficiaries to join the cooperative. The promoter's perception is conditioned by a number of factors including the image he has of the dynamics of a peasant community and his expectations of the behavior of a "key" person. His attitudes toward peasantry, mestizo middle holders, and the agrarian reform also enter into his perception.

On the other hand, the peasant community has an image of the type of person who can best fill the role. Ideally, such a person would have resided in the community for some time, would have indicated his allegiance to locally prescribed codes of behavior, and would have held important positions in the civil-religious hierarchy. The *personero* office in independent communities has traditionally corresponded to the type of person selected for spokesman.

The dilemma that emerged at this stage of the selection process is that there are few peasants qualified to interact in the larger society who at the same time have all the requirements from the perspective of the local community. At the least, an individual must be literate, fluent in Spanish, and able to interact within the institutional setting of the larger society. As we have seen in Chapter 2, these traits usually select out those who are most capable of living in the city and who eventually migrate. The pool is thus relatively small in relation to the population of adult peasants. There are some peasants who have returned from a sojourn in the city or service in the military. Usually, their attitudes and their experiences, which are primarily "urban" oriented, have neglected participation in the civil-religious hierarchy as a source of prestige meaningful to them. Thus, while a peasant may well be qualified to act out a broker role, there is no guarantee that he will act in the best interests of the community. This explains in part why in independent communities there is both respect and fear of the *misti* stratum that has recently emerged from among the ranks of the puna peasants.

The situation is especially critical in ex-hacienda communities like Tukiwasi. Since the hacienda and the supra-household sphere of the peasant economy is run by the administrator of the hacienda, there is no autonomous decisionmaking body made up of *feudatarios*, and thus no channels for the emergence of talented leadership. Social controls and the nature of the patron-client relations act to inhibit the availability of leaders so crucial in the transition to local autonomy.

The role of spokesman is a niche in the community that an individual may use for other ends notably the aggrandizement of power. As Barth notes (1963:7), such a newly created niche in the set of social relations pattern of a community is an optimal location for an entrepreneur; it enables one through judicious transactions to add to his store of wealth, whether it includes factors such as power, rank, experience, and money.

The performance of an individual in transmitting information is only partly related to the process by which he was selected for the role. It is constrained first by the extent to which he meets the qualifications of an ideal spokesman from the perspective of the community and, secondly, by the extent to which his performance is perceived to act in the best interests of the community. His performance in this regard can vary considerably. If a discrepancy exists between an individual's background and the ideal as perceived by the community, the uncertainty may increase. Under these circumstances, a rational decision would be to wait and observe the subsequent performance before a commitment is made. This appears to have been the case in Rumipata. If, on the other hand, the spokesman meets the requirements of experience and responsibility to the community, then uncertainty is reduced. Under these conditions, the dilemma of the peasant is reduced because he can rely on the past performance of the individual to continue into the future. Both Tukiwasi and Antapampa meet these conditions; the major difference is the motivation of each person: RP in Antapampa, in contrast with MCS in Tukiwasi, did not directly attempt to negotiate with the agrarian reform bureaucracy as did MCS in Tukiwasi. His motivation in other words was sufficient enough to seek and accept the broker role but not strong enough to represent the community's interest.

8

Conclusion

The analysis of the impact of the agrarian reform in the Pampa de Anta region of southern Peru shows that it has been successful in removing the basis of power—control over land—that an agrarian elite successfully utilized in maintaining its position in the political economy.[1] This process, accomplished through the expropriation and redistribution of land and the isolation of the agrarian elite from other sources of power such as the juridical, political, and communications institutions, has brought about considerable structural change. In the Pampa de Anta, the agrarian elite never directly dealt with events, except during the period until the last two decades or so, but rather allocated power to a dependent mestizo class that represented its interests. Now, following the agrarian reform, the support that the mestizo class previously relied upon has been removed.

In place of the agrarian elite and its dependent mestizo class, there has emerged an agrobureaucracy that has sought to control the institutions usually associated with the agrarian elite. Inasmuch as the agrobureaucracy is directly linked to Lima—the national capital—there has been a shift in the locus of power from the regional to the national level.

In one sense, then, the goal of the agrarian reform in affecting significant changes in man-land relations has been met. In those highland regions such as the Department of Cuzco major readjustments will have to be made to the new situation including changes in the relationships of social groups in the urban centers.

When we turn from changes in landownership and the composition of the elite to those goals targeted specifically

1. This chapter is a substantially revised and updated version of an article previously published in the *Journal of Rural Cooperation* (Spring 1978).

for the peasant beneficiaries of the agrarian reform, the situation is far different. Our data show that participation in the cooperative Túpac Amaru II, whether it be measured by membership, input into decisionmaking, or sharing in an increase in production, is less than optimum. Here we will review the factors that lead to this conclusion and examine the extent to which they are found in other agrarian reform enterprises in Peru.

The agrarian reform emerged out of a series of events at the national level, particularly as a response to changes in the social structure of coastal society. The military government sought to redress the societal imbalances that had plagued its agrarian sector and, at the same time, sought to create a dynamic and progressive vehicle through agrarian reform that would bring the agrarian sector into the national economy. The lack of opposition at the national level is noteworthy and contrasts with earlier attempts at agrarian reform in Peru, which were usually emasculated by legislative and administrative maneuvering.

At the regional level, in Cuzco and Pampa de Anta, we find a different set of actors more typical of highland Peru. In Cuzco, which is the urban and administrative center of the region, opposition failed to materialize from the traditional agrarian elite. The reasons, though, are different from those that account for a lack of opposition at the national level. Several factors, including a turning away from a rural land-based lifestyle, the peasant movement of the 1960s, and fractionalization of properties through inheritance, had weakened the motivation of the agrarian elite and its ability to mount an effective opposition. We know that, in some instances, personal ties between landowners and government officials did result in one's property passed over or set back in the timetable for expropriation. However, this was the exception rather than the rule. Generally, the government was able to expropriate the major holdings in the Pampa de Anta with relative ease. The major obstacle was a result of an understaffed and inefficient bureaucracy. But this is perhaps understandable: In the haste to organize an effective agrarian reform program in the region, there were problems in recruiting skilled and trained manpower, especially given similar demands in other zones throughout the country.

Communication Flow in the Pampa de Anta

In the implementation of the agrarian reform in the Pampa de Anta, the government was faced with a difficult and more subtle problem than the quelling of opposition from an entrenched elite. The problem stemmed from the mode of distribution that was chosen. Unlike other cooperatives in Peru formed during the agrarian reform, in the Pampa de Anta peasant beneficiaries were recruited into the cooperative Túpac Amaru II as individuals, not incorporated as communities as in the SAIS units, or as entire functioning enterprises as in the sugar complexes of the coast. The government was thus forced to "sell" the cooperative to peasant beneficiaries in order to motivate them to join and to participate.

To acquaint the beneficiaries with the cooperative and to inspire them to become members, the government created a communications program that is modeled on the communication of innovations approach to social change. The data in this study indicate that this program has been largely ineffectual in achieving its goal.

Do the causes of the failure of the communications program lie in the nature of the program itself or in the population of peasant beneficiaries?

The Pampa de Anta is admittedly a difficult region in which to implement such a program. Roads are poor, peasant communities are widely dispersed, telephone and telegraph systems are only rudimentary, and peasants are often away from their homes most of the daylight hours. Despite these obstacles, the government program appears to have been well planned and adequately funded. It greatly emphasized the "oral" media channels, such as mobile public address systems and the Radio Forum program, which could be transmitted in Quechua to the essentially illiterate peasant population.

The content of the communications program was extremely complex and only partly capable of being transmitted orally. Statements of agrarian policy have been disseminated in the form of laws, annexes, and legal documents. Most of these statements are difficult for even a relatively educated reader to follow: They include complex legalistic phrasing, numerous rules and regulations, exceptions to the general rule, references to preceding legislation, and other features common to legal documents. An individual wishing to be fully informed would consult a lawyer to assist in interpreting the content of a document. Obviously, a peasant does not have these resources. Although an independent judge was appointed to hear cases involving the land questions, it is fair to say that there has been general confusion as to the interpretation of major agrarian legislation.

Part of the confusion stems from difficulties in bicultural communication. The mass of the peasants in the southern highlands are illiterate, often monolingual, and have little experience dealing with the bureaucracies responsible for the administration of reform laws. Even translation of these documents into Quechua, which was requested by peasant representatives early on in the agarian reform, would not have solved the problems. As was said earlier, since Quechua is not a written language, translation into Quechua would require a phonetic translation that would encumber its usefulness. Obviously, such a translation would not be really instrumental in helping Quechua-speaking peasants who did not read to begin with.

In lieu of an effective communications program, our data show that individuals who act as informational links between the local level and the larger society have been instrumental in communicating the agrarian reform. This role or set of roles was first referred to as the "broker" by Wolf (1956:-1075) to mean those individuals who "stand guard over the critical junctures or synapses of relationships which connect the local system to the larger whole." According to the types of links that are significant in any given situation, there may be varying degrees of institutionalized broker roles in society. Here I am discussing what Schaedel (1972) has referred to as the culture informational link between the community and the macro-society; it refers to the role the broker plays in the flow of information between societal segments that bound culturally distinct systems.

The role of brokers in the perception by peasants of economic alternatives coming from the larger society is based on their function of acting as transformational agents in interpreting economic alternatives in the languages and conceptual systems of the peasants. Since his role stems from the real or putative ability to act within the cultural and social systems of both the local community and the larger society, the broker has access to knowledge that he transforms into readily understandable elements that may then be internalized by individual peasants. In this sense, the broker performs a different function than the "opinion leader," which according to the "two step hypothesis" of the diffusion of innovation literature is influential because he either provides a "relay" function, relaying public information to the previously uninformed or a "reinforcement function," increasing the clarity or effectiveness of some specific public information by repeating it to already exposed group members (Allan Pred 1967:-35–36).

There is a second aspect of the role of the culture-informational broker that is relevant here. In societies like Peru with a large, characteristically illiterate, indigenous population, the ability to manipulate information originating in the larger society can be an important source of power for the broker operating at the local level. In southern Peru, in particular, control over information has been one of the primary means by which mestizos have maintained their superordinate position with respect to the peasantry. This aspect of the role of individuals involved in the transmission of information is not adequately acknowledged in the diffusion of innovation literature.

A moment's reflection might reveal some of the reasons why information as a source of power has been missing from the diffusion of innovation literature. To a large degree, this school of applied social science is based upon research in directed change conducted in industrialized societies. For example, probably the most common type of situation encountered in this literature is the role of the extension agent in introducing a new type of seed or agricultural practice. This process is admittedly analogous to the problems encountered in the Third World, notably among peasant populations, where similar traits are disseminated daily through both diffusion and directed cultural change. In the diffusion of innovation literature, much discussion has revolved around the role played by links such as the extension agent in the flow of information. The role of extension agent, however, differs in many respects from the culture-informational broker linking the peasant community with the larger society. The extension agent performs his job essentially in order to earn his salary. The culture-informational broker on the other hand is a non-institutionalized role that an individual may aspire to out of an interest in the power that might accrue. In so doing, he has access to a much greater range of power than does the extension agent; this develops from the nature of traditional societies undergoing "modernization." These societies are often marked by relations between peasant and the larger society that Richard N. Adams (1970) has referred to as a domain structure—a relationship in which one set of actors has greater control over another set than the latter has over the former. In the domain structure, brokers usually are instrumental in the exchange of power. Such situations can be contrasted with one in which domains are weak and levels of articulation between actors are flexible. In this case, access to power is facilitated by "career mobility systems."

In societies in which domain structures prevail, information may be an important source of power manipulated by superordinates in the domain relationship. Information often is transmitted only by individuals meeting certain requirements, usually educational, linguistic, and attitudinal in nature. Certain of these factors are in short supply in peasantry. As a result the ability to impart information becomes a crucial source of power both sought for and competed over by groups attempting to optimize their location in the power structure.

Our data indicated that the politicization of information has influenced considerably the perception of the agrarian reform by peasant beneficiaries. As a complement to the formal communications program, the government attempted to utilize informal linkages analogous to the role of mestizos in the communications network of the pre-agrarian reform political economy. Their reasons for doing so were twofold. They were interested in communicating their program to the peasant beneficiaries and recognized the importance of informal, personalistic, sources of information, and they were concerned with peasant reaction, especially of a collective nature, to the agrarian reform since it had the possibility of releasing tensions that had built up during the pre-reform period. The role model that was employed was that of the "peasant leader" who had emerged during the peasant unrest of the 1960s. As we have seen in our analysis of EP, the dilemma was to locate legitimate peasant leaders—individuals who were capable of interacting in both the local level peasant communities and the larger society and who were aware of the represented peasant interests—in a fluid situation ripe for the machinations of political entrepreneurs. As the case of EP demonstrates, some "peasant leaders" who emerged during the selection process were guided by narrow, self-serving interests and had at best only localized peasant followings. The image of the agrarian reform at the local level has suffered through the selection and performance of some of the "peasant leaders" associated with it. This kind of newly established linkage did not remove the influence of mestizos on the local level but merely bypassed it. Mestizos have continued to interpret the agrarian reform to peasants in a bad light, further confusing an already difficult situation.

In the Pampa de Anta following the 1969 agrarian reform, the ability to process information became a resource in competition in the local level political arenas. The clauses prohibiting indirect usufruct caused the return of a number of outmigrants who had used this means to control their land. On return, the migrant was faced with readapting to life in the community and solving an "identity" crisis resulting from his perceived downward mobility. Our data from Rumipata indicates that migrants have been active in the ritual and political spheres of community life. These spheres are optimum locations in the local social structure where migrants can best manipulate the resources at their command, including a knowledge of "urban" culture and institutions, national language, and educational skills, and the reputation that derives from having lived in the city.

Other groups demonstrated these same qualifications. Mestizo middle holders were particularly important since they felt threatened by the agrarian reform and sought for the first time to enter into local level politics in order to establish themselves beyond the reach of the agrobureaucracy. The competition that ensued did not always pit mestizo middle holders against return migrants; each community had a different mix of these and other groups. But, given the nature of post-agrarian reform change, individuals and groups who have been able to wield these resources were extremely active in the political processes, and the role of brokerage, at the local level. Their performance in establishing linkages between the peasant community and the cooperative, has determined, to a large degree, the perception of the agrarian reform by the peasant beneficiaries, and ultimately its success in recruiting and maintaining an active cooperative membership. In Tukiwasi, MCS was brought into the formal communications network as an accommodation by the government to his position in the hacienda community before the agrarian reform. MCS was thus able to negotiate on behalf of Tukiwasino interests and, in effect, insulate them from direct interaction with the cooperative. Tukiwasi residents perceived the cooperative that emerged as merely another kind of hacienda in which MCS continued to act as patron in an ongoing patron-client relationship. In the other communities, peasants were not so fortunate. Through the influence of several situational variables, described in Chapter 5, the performance ranged from the conflict-free transition of Tukiwasi to the confusion and factionalism of Rumipata.

If the planners could have resolved the problems of communication in a bilingual-bicultural setting, and if they could have removed the threat that politicization of information presented, would their program have been more successful? Charles Erasmus, an anthropologist who has worked in various projects of applied anthropology, has constructed a theory of culture change (Erasmus 1961), which would seem to bear on the problem. Culture change, according to Erasmus, is based on "frequency interpretations" or the "potentiality to synthesize experiences into predictive generalizations" (p. 31). Knowledge follows from frequency interpretations and their observed probability of reoccurrence. Frequency interpretations, then, involve, essentially, the reduction of uncertainty and the translation of the decision concerning a culture alternative to risk. To Erasmus, there is a basic difference in the way knowledge is obtained in different societies: "In literate industrial societies, frequency interpretations are revised largely by technical observations of specialists who have greatly accelerated the rate at which refinement of correlations take place" (p. 31), while "cognition among uneducation, unspecialized, people is based largely on direct experience" (p. 22).

It would follow from Erasmus's theory that in a complex cultural alternative, which the cooperative represents to the beneficiaries, some aspects would be observable and thus provide a base for "experience" and others would not. An evaluation of the merits of the alternative then would be heavily weighted in the results that observation yielded.

Our data suggest that peasant beneficiaries, especially from independent communities, have evaluated the economic performance of the cooperative based on observation of ex-hacienda land expropriated by the agrarian reform and cultivated by the cooperative. These lands are adjacent to independent communities and have provided a "laboratory" through which cooperative agricultural operations could be weighed against the peasant's own background of experience, knowledge of the local ecological setting, and awareness of successful agricultural practices. Their observations indicated a high degree of uncertainty in the long-term economic performance of the cooperative.

There were several factors that led to this conclusion. They were well aware that the success of agricultural or cattle operations in the Pampa de Anta depended on rapid responses to the vagaries of seasonal rainfall, temperature, and other climatic variables. If seed, fertilizer, equipment, and labor are not made available at the correct time, then the entire crop is jeopardized. Yet, during the period when the field study was made, it was clear that the cooperative lacked the flexibility, speed, and planning to respond adequately to production needs.

Part of the uncertainty, from the perspective of the peasant beneficiary, has been due to the changes in social relations brought about by the agrarian reform. Cooperative administrators and advisers were often recruited from outside the region and, in many cases, were graduates of agricultural colleges on the coast. Certainly, there were valid reasons behind recruitment practices: Experienced personnel were scarce, and there was an effort to bring in individuals who would not be compromised by conflicts of interest. But peasant beneficiaries were quite suspicious of these administrators who stressed social distance and were unaware of the role that social change could have in their performance as economic advisers. As has been reported for other areas of Peru (Rodrigo Montoya et al. 1974:89), agrarian reform advisers and technicians see their role as maintaining pre-reform production levels and, if possible, raising them. These essentially economic goals are felt to be more important than the "social" ends of the reform. To the technicians, social and political questions were not their responsibility. This situation exacerbated the negative image peasants had of their performance. Even their questions designed to explore the range of local agricultural practices were misunderstood by peasants as an indication of incompetency. In sum, from the perspective of the beneficiaries, risk was increased by the nature of the changes in social relations that the agrarian reform hoped to bring about.

Both the cattle operations and the agricultural operations of the cooperative have not adequately met the needs of the range of peasant interests in the region. *Yerbajero* peasants, seeking access to pasture, discovered that ex-hacienda pasture lands were to be plowed up for potato cultivation and those lands that were finally set aside for pasture were prime locations for cattle thieves who acted with impunity. Opportunity for work on cooperative lands never materialized since the periods of peak labor need on the cooperative corresponded to the same peak labor periods in the households of peasant families. The cooperative did not ameliorate the situation by investing heavily in labor-saving machinery. Investment in the cooperative was viewed as quite risky, given limited experience with the cultivation of ex-hacienda lands. In short, when we look at those specific "income streams" that would have been appealing to peasant beneficiaries, we find that in most cases they failed to materialize.

Structural Incongruity

While direct observation proved useful in enabling peasants to evaluate the economic performance of the cooperative, it was less effective in communicating the nature of the cooperative as a model of production. One reason is that it is much more difficult to observe the functioning of a sophisticated behavioral system than the cultivation of a crop on a given plot of land. Another, somewhat deeper, reason was the structural incongruity between the organization of production in the cooperative and the organization of production in the peasant economy. This incongruity is most pronounced in the economy of the independent community and is illustrated in Figure 8-1.

Figure 8–1. Loci of Production Decisions

	Community	Cooperative
Short Term	Household	Delegate Council
Long Term	Supra-Household	Administrator

The incompatibility lies at two levels: at the loci of short-term and long-term production decisions. Peasants operate household economic units and make decisions allocating factors of production at their disposal. In the cooperative, on the other hand, short-term production decisions are a responsibility of an individual or set of individuals delegated authority by the upper levels of the cooperative. As such, members of the cooperative do not have any input into short-term production decisions.

Long-term decisions in the community are made either in the assembly, in independent communities, or by the *mayordomo* in hacienda communities. In independent communities, they emerge through the tedious process of group decision-making. This process is based on normative rules that include face-to-face discussion of issues and agreement through consensus. Anybody who has something to say about an issue has the right and is expected to speak out before the others and present his case. Once a decision is reached through consensus, sanctions, both formal and informal, act to ensure that all peasants abide by it. Long-term production decisions in the household are concerned with reciprocity, redistribution, and renewability of resources; they are oriented to the viability of the individual household over time. Long-term policy does not usually involve the collective exploitation of resources held by the community; rather it is concerned with allocating resources to households on an individual basis. As Georgescu-Roegen points out (1969:73), redistribution in peasant societies is done in order to equalize the opportunity of achieving household production rather than equalizing income. While there is a sort of social welfare that obtains in Andean communities, close examination reveals that it is concerned with the disadvantaged: orphans, the elderly, widows, and the like. Georgescu-Roegen's comment regarding redistribution could be extended to the other functions of the supra-household: reciprocity and renewability of resources. Their rationale is to maintain what Chayanov (Thorner, Kerblay, Smith 1966) calls the "constant state of well-being" of the individual household. To view the supra-household as a joint shareholding enterprise in which resources are exploited in common would be a serious mistake.

Long-term production decisions in the cooperative are based

on collective exploitation of jointly held resources in a sophisticated production unit. They are based on the maximization of productivity of the cooperative as a unit with the expectation that rewards will be shared among all of the members contributing to its production. It necessarily is complex; an involved hierarchical division of labor, a sophisticated, codified, set of rules and regulations regarding its operation and the delegation of responsibility are some of its characteristics. Most importantly, though, the long-term production decisions are made through a much different process than in the peasant economy; they are made through delegates elected from among nuclei of cooperative members who represent their constituency in a general assembly. Thus, local members neither participate directly in long-term production decisions unless they are elected a delegate nor in short-term production decisions, which are the responsibility of an individual or individuals delegated authority by the upper level of the cooperative. The cooperative stands then as a production unit with significant differences in the formulation of long-term policy: differences in terms of the orientation of long-term production decisions and differences in the manner in which they are made. In effect, it approximates the model of the agrobusiness, oriented to the maximization of production and profits.

In retrospect, the agrarian reform planners fell victim to a common misunderstanding of the nature of the peasant economy in the central Andes. This misunderstanding derives from a veneer of cooperation and communalism in rural settlements. For example, reciprocal labor between households and faena group labor within a collectivity of households is certainly common, but not universal. And there is a pattern, although usually truncated today, of collective control over land, most often in the puna ecological zones. These patterns have led many observers to argue that there is an underlying tendency toward cooperation and communalism in the peasant economy, and, further, that this tendency is an important resource that can be drawn upon in the design and implementation of socioeconomic development strategies. The earliest exponents of this view are the "Indianista" writers, José Carlos Mariátegui (1967) and Hildebrando Castro Pozo (1924). In the 1950s and 1960s, it underlay the programs of community development then popular with national and international development agencies. In the writings of the "Indianistas" and the community development literature, the community is seen as a cohesive, "natural" social unit, with a set of characteristics leading to its perpetuation in the face of threats from the larger society. It represents a storehouse of energy that can be tapped for development and fosters social relationships that are cooperative, helpful, generous, and "boy scout like" (Adams 1962:409–10). The eclipse of this development strategy occurred for two reasons. First, for a number of reasons self-help was difficult to promote, and the programs failed to significantly increase the well-being of peasants. Second, there was a contradictory body of literature emerging in the same period in anthropology which held that peasants were essentially suspicious, envious, and uncooperative (Oscar Lewis 1951); some went beyond description to argue that there was a basic cognitive set, the famous zero-sum game model, to peasant personality that militated against cooperation in the collective good (George Foster 1965). Although subsequent research has shed doubt on the validity of the zero-sum game model of peasant behavior (Kaplan and Saler 1966, Kennedy 1966, Bennett 1966, Piker 1966, Acheson 1972, Gregory 1975), it continues to surface in the explanations of planners and social scientists for the failure of peasants to respond to community development programs and other programs of directed change (Rogers 1969). One of the legacies of this debate is that the search for a common character to peasant social relations, in particular, the determination of whether or not they reflect "tendencies" to cooperation, is doomed to failure. Social relations are best seen as an outcome of the production process. In each empirical case, these relations are determined by (1) production constraints on individual households that call for cooperative or collective action, and (2) the long-term production strategy of a collectivity of peasant households interacting in a natural settlement (Guillet 1978*a*).

Since the late 1960s, there has been a return to the organization of production as a development strategy as the Green Revolution proved to involve too many obstacles. In this latest reincarnation of the earlier community development approach, existing small peasant holdings are consolidated into more efficient, larger size and scale, production units, and haciendas that are expropriated or purchased and kept intact to be farmed by a group of peasants. This basic strategy has many variations but is the basic approach now taken in Colombia (the *empresa comunitaria*), Panama (the *asentamiento*), Venezuela (the *empresa campesina*), and, of course, Peru.

Centralization, Size, and Scale

It should be apparent that centralization of power and decisionmaking has characterized the formation and subsequent operations of the Túpac Amaru II cooperative. This centralization is manifested in several ways. Major decisions concerning the value, manner of expropriation, and mode of distribution of haciendas were made by a commission of government advisers in Cuzco. Their recommendations proved to be heavily dependent on decisions taken at the national level. Beneficiaries were qualified by SINAMOS. Long-term production plans are adjusted to goals set by the Ministry of Agriculture; they largely reflect the needs and demands of urban consumers and voters. During 1971–1972, SINAMOS exercised considerable control over the political trajectory of the cooperative and intervened in the resolution of labor disputes. Upper level advisory and administrative positions are proposed by the Agrarian Reform Agency. The government-owned Agrarian Bank is the major source of both long-term and short-term credit for the cooperative. Salary increases are agreed on at the highest levels.

Certainly there are arguments that can be mustered in favor of centralization. The Latin American experience, particularly the Mexican and Bolivian agrarian reforms, shows that failure to adequately create a long-range plan based on an analysis of the existing socioeconomic structure and a target to which policy is oriented can imperil the ultimate success of a program. Centralization, it could similarly be argued, is necessary in order to facilitate quick decisions on policy alternatives and to efficiently utilize a limited pool of qualified administrative and technical personnel. Nevertheless, it has led to a lack of flexibility and feedback, which has jeopardized production and productivity and severely constrained participation. As a result, while some power has been transferred to beneficiaries, it consists basically of the ability to make relatively minor decisions rather than the full participation in a self-managed enterprise as the government rhetoric of agrarian reform promised.[2]

The problems implicit in centralization were increased by the size and scale of operations chosen for the cooperative; they in turn are related to two important early decisions. First, there was a shift very early in the agrarian reform program from a scheme of "land to the tiller" to one of production cooperatives. Second, somewhat later, a decision was made to create in the Pampa de Anta one rather than three or four separate production cooperatives. The former placed demands on the credibility of community representatives who had to report the shift in policy to their communities and it resulted in a wide sense of disappointment. The latter was a result of the desire to maximize the impact of the cooperative as an example of what the agrarian reform could accomplish in the southern highlands. The incorporation of all the expropriated haciendas and the recruiting of beneficiaries from among the adjoining communities into one large production unit had the effect of forcing what were often antagonistic social groups into "cooperating" in production. One must not forget that Andean rural history is replete with longstanding antagonisms among communities and between communities and haciendas based on disputes over land. By this action, preexisting tensions and conflicts were incorporated into a cooperative unit that must be based, for it to be successful, on a sense of trust between members and a sense of loyalty to the group (Ronald F. Dore 1971).

The large size particularly exacerbates the inherent problems of centralization. First, it makes more difficult the resolution of the preexisting sources of tension and conflict. Second, it increases the problem of articulating the goals of the cooperative in such a way that they can create a consensus and a sense of loyalty among its membership. And, lastly, it creates obstacles to the generation of a long-range production plan that must take into account a range of local practices, interest groups, ecological zones, and community structures.

It would appear that size alone is one of the most significant problems associated with the success of the production co-

2. For an insightful analysis of the limited range of decisionmaking available to beneficiaries see Sean Conlin 1974.

operatives in the southern highlands. Horton, for example, in a review of several of these units stated:

> In all cases observed (in the Department of Cuzco) membership participation in the implementation of reform and in the management of reform enterprises is negatively related to enterprise size. In many cases there is a trade-off between productivity and participation. In Santa Lucia, for example, fusion of several ex-estates has seemed to help boost livestock production, but members participate only marginally in management. In extreme cases, such as the (production cooperative) Tupac Amaru II large size reduces both productivity and participation. The large "model land reform enterprise," designed by outsiders unfamiliar with practical aspects of crop and livestock production, is simply too large to manage. (Horton 1974:11.13)

In retrospect, it would have been much more advisable to have created small production units that more clearly corresponded to sets of naturally interacting haciendas and independent communities in the Pampa. As an example of the results that might have been achieved by this strategy, the successful case of the transformation of a hacienda in Ayacucho, a neighboring region in the southern highlands, comes to mind (Susan C. Bourque and David Scott Palmer 1975:206–8). In this instance, size and scale were kept low, with a total of 515 hectares and thirty-seven families, and the locally relevant ecological zones were included in a mixed production strategy. The zones included high altitude pasture, puna land optimum for tuber cultivation, and fertile irrigated bottomland, which had been the property of the hacienda. Following the creation of the cooperative, peasants continue to retain their individual plots in the tuber zone, have collective rights to pasture and firewood in the highest zone, but collectively farm the irrigated bottomlands. The ingredients of this strategy thus obey the fundamental principle of verticality, combine both subsistence and cash cropping to reduce risk, and restrict market production to the fertile bottomlands.

Agrarian Reform and Cooperatives in Other Regions in Peru

One might legitimately ask the question "To what extent do the patterns with respect to participation in the cooperative Túpac Amaru II occur in agrarian reform cooperatives in other regions of Peru?" It is difficult to answer the question in its entirety; studies of other cooperatives are limited and do not ask the same questions asked here. Further, this study has stressed the options, and the associated constraints, involved in peasants' decisions to join and participate in the Túpac Amaru II cooperative. Other cooperatives have either incorporated entire functioning production units, such as the coastal sugar cooperatives, or production units and adjoining peasant communities, as in the SAIS cattle operations in the central highlands. However, participation in each type of cooperative and its effects on economic performance are subject to some of the same factors that have been isolated in this study. In order to provide some insight into the operations of these other types of cooperatives, we have chosen two: Pomalca, a large coastal sugar cooperative and the SAIS Cahuide in the central highlands.[3]

Pomalca is an enormously large and complex agro-industry, based on the cultivation and processing of sugar cane, located in the Lambayeque Valley of Peru's north coast. It comprises a total area of 16,796 hectares and a labor force of 3,249. It is occupationally diverse, including a stratum of administrators and technicians; office workers and other service personnel, salaried workers engaged in day-to-day field and factory operations; full-time resident laborers; and temporary workers.

As in any large essentially mechanized enterprise, occupational diversification breeds tensions between the various hierarchical rankings. These tensions existed in Pomalca before its formation into a cooperative but were kept under the surface by the upper levels of administration. With the shift to a modified form of workers' control following the agrarian reform, the tensions emerged in December 1971 in a strike over the performance of four technicians. The issues included a feeling of distrust among the workers toward these technicians, their higher salaries relative to field and factory workers' wages, and their ties with the ex-hacienda owners. They were ultimately expelled as a resolution to the problems.

"Vertical" communication flow, between the field and factory workers and the resident and temporary laborers, and the governing bodies of the cooperative, is beset with problems

3. Material for the section on Pomalca comes from: Douglas E. Horton 1973; CENCIRA 1973; and Colin Harding 1974. For Cahuide: B. Roberts and C. Samaniego n.d.; Ron Skeldon 1974; Rodrigo Montoya et al. 1974; and CENCIRA 1973.

similar to those encountered in the cooperative Túpac Amaru II. Ideally, it is based on delegates who transmit decisions on cooperative policy to their constituencies and, in turn, articulate lower level interests to administrative decisionmakers. There are problems, though, in the flow of information. Delegates often fail to receive messages and misinterpret them in their reports to the local nuclei of cooperative members. The formal media program does not appear to be of much use: "field workers fail to grasp written messages (bulletins, communications, etc.) and the use of mobile units as a means of mass communications is essentially inefficient in the countryside. High level cooperative officials are only sporadically in the fields, with the result that the distance between the conception field workers have of the cooperative and reality is great" (CENCIRA 1973:10).

Obstacles to communication extend to groups that are at approximately the same position on the hierarchical occupational ranking. Cooperative members are not generally aware of issues of interest to members of groups other than their own. There are exceptions, to be sure. Some members of occupations, such as machine operators and factory service personnel, come into contact with one another since their tasks locate them in the same place in the cooperative. Often issues such as wage scales, working conditions, and overtime policy affect the interests of several groups. However, lack of effective horizontal communication confuses the perception of common interests and hinders the consensus needed to press for resolution of issues in the governing bodies of the cooperative.

Despite these obstacles to the flow of communication, Pomalca appears to be an efficient and smoothly operating enterprise, especially in comparison with Túpac Amaru II. Part of its success can be attributed to centralization: All major technical, financial, and administrative decisions are reviewed by the Central Sugar Cooperative, an umbrella organization that oversees the operations of all the coastal sugar cooperatives. The administrative council of Pomalca, which ordinarily would make these crucial decisions, spends most of its time on questions of a social nature (Horton 1973:64).

Perhaps the most relevant measure of economic performance is that Pomalca has not only met pre-reform levels of sugar production but also has shown increases. Total cash income in the cooperative has grown, and field workers, the most underpaid group during the hacienda period, have been granted greater pay increases (including subsidized rations, medicine, and housing improvements) relative to other occupational groups. There are indications, though, that sharing among workers in increases in productivity will require adjustments in the labor force. Cooperative labor has become far more expensive than temporary labor, suggesting that "the number of cooperative members will fall (through death and retirement) while the ratios of outside labor to cooperative labor and capital to labor rise" (Horton 1973:72).

The role of temporary labor has been a crucial issue in the evolution of the agrarian reform on the Peruvian coast. Most of the estates that were formed into cooperatives were based on temporary labor since it was less expensive than permanent resident labor. Cooperatives, though, have been run as strictly commercial enterprises with their capacity to absorb labor dependent on their productivity. Their inability to absorb temporary laborers resulted in militant demands for their inclusion, while those permanent laborers who were included have been determined to keep them out. The government's response has been to form larger production units on a regional basis. One analyst has argued that this trend to the creation of large-scale production units goes against the original goal of the agrarian reform to encourage efficiently run private farms (Harding 1974:15).

Cahuide is located in the Junín Department of the central highlands. It is a social interest society (SAIS) based on a large-scale cattle operation. There are two separate sets of components of Cahuide. The first is a large cattle hacienda, the Sociedad Ganadero del Centro. Its land and permanent work force, including herders, laborers, service personnel, and technicians, have been organized into a cooperative and incorporated as the main production unit of SAIS. The second is composed of twenty-nine separate highland peasant communities that were felt, after an analysis of the situation, to have had significant relations in the form of labor recruitment and pasture rights on the adjoining hacienda. Each of these twenty-nine communities has been incorporated into Cahuide as an autonomous unit. Both component parts come together in the delegate assembly, which is ostensibly the main governing body of SAIS.

Cahuide has had many of the problems found in Túpac Amaru II and has not been able to achieve the relative harmony and efficiency of Pomalca. While centralization and a "rationalized" approach to production reaped rewards in Pomalca from the standpoint of economic performance, it has resulted in Cahuide in the familiar conflict between the interests of the advisers of the unit and those of the component parts, in this case the production unit and the highland communities. Administrators identify with the economic goals of the SAIS which they interpret as the maintenance of production and, if possible, the increase of production levels. An overconcern with these narrowly drawn goals has created obstacles to participation and the broader picture of agrarian economics.

There are numerous symptoms of the overemphasis on centralization and rationalization. There has been considerable resistance to representation of collective interests of the workers of the production unit. In the delegate assembly, the production unit is represented by only two delegates while each highland community sends two delegates for a total of fifty-nine. This discrepancy has been a source of much tension, especially since the interests of the communities differ from those of the production unit. Following the loss of a crucial vote by the production unit in a meeting in December 1971 of the delegate assembly, there was a movement to create a union. They were resisted by the administrators of the production unit (B. Roberts and C. Samaniego n.d. 13; Montoya et al. 1974:58). This is an attitude that is characteristic of administrators, a lack of interest in general in the political participation of SAIS members.

Another sign is the pressure put upon highland communities to adopt a more efficient cooperative organization, or at the least to engage in "cooperative" projects to develop their socioeconomic infrastructure. This pressure on the highland communities is based on a similar misconception that was observed in the Pampa de Anta; a belief that the cooperative mode of organization "fits" the socioeconomic organization of independent communities. In the region of Cahuide, though, there is even less of a basis for such an organization. According to a study by Roberts and Samaniego, highland villages were formed as a consequence of the need for an interchange of products from various "vertical" ecological zones. Shepherds entered into contractual relations with households in the lowland zones to take care of their herds in the puna. Over time the shepherds began to accumulate herds as a result of the contract that gave them part of the offspring as payment for herding. Complex interdependencies developed between the lowlanders and the herders and the herding settlements crystalized into permanent villages. Later a complex and multifold process brought further internal diversification in the highland villages; the hacienda provided opportunities for wage labor as did regional mining enterprises, coastal plantations, and urban and rural public works. The richer stratum of the communities expanded their herds; when pastureland became scarce, they intensified their search for capital investments, mainly in land, in the lowland villages. Contemporary highland villages reflect these historical processes: Members of the communities are economically diversified, with production strategies that are multilocal. Along with diversification, stratification has created a series of vertically ranked groups.

Economic diversification and stratification of relatively acculturated highland communities has created major obstacles to their participation in the SAIS. Besides a strong household orientation, there is no shared pastoral subsistence strategy but rather a variety of strategies that are multilocal in nature. Most of the interest in the SAIS has come from the richer and more acculturated stratum of the community:

> Our data show that it is these larger farmers that are most interested in and know most about the SAIS and its organization. The majority of members of a community evidence little interest in or knowledge of the SAIS. According to the questionnaire conducted by the SAIS Cahuide development division 49% of the community members had no knowledge or understanding of what the SAIS was; a further substantial proportion had little understanding. (Roberts and Samaniego n.d.:14)

Presumably the situation described, which is similar to that obtained in Rumipata and the wealthier communities of the Pampa de Anta, contrasts with the lack of participation of the community members who could theoretically most benefit from the SAIS. But the "rational" capital intensive policies of the administrators have failed to provide work for those members of the highland communities who would most benefit from it. There have been belated plans in SAIS units in the central highlands to create a demand for labor by colonizing the adjacent tropical regions, but there is evidence that it is insufficient (Skeldon 1974).

Cahuide, because of the expanse of sparsely settled pasture land that it encompasses, is particularly susceptible to obstacles in communication flow like those noted for Túpac Amaru II. Day-to-day directions are given orally by administrators to managers of local production units. Long-term policy emanates from the advisory council where it is transmitted to delegates and through them to local nuclei of workers. Most SAIS members live at a considerable distance from the central offices, though, and messages arrive late, if at all. Delegates, particularly from the more isolated communities, are illiterate and must rely on memory to communicate complex and important policy decisions. When delegates are not available, because of health or other reasons, replacements or the use of the mails have proven totally inadequate. Given these and other factors, rumors are endemic and no real mechanisms in the flow of information exist to clarify them.

The reverse flow of information from local assemblies and production units to higher levels of the cooperative is highly inefficient. Demands are presented to administrators rather than through the formal channels set up for adjudication. Herders, the most isolated and dispersed of all the groups, are perhaps the least able to meet SAIS representatives to seek redress or give input. In sum, communication flow in Cahuide is constrained by a breakdown of linkages at the local level and the effect of distance.

Conclusion

The agrarian reform in the Pampa de Anta illustrates the classic conflict between the "national" goals of a government in designing agrarian policy and the interests of a peasantry who are the beneficiaries of the policy. To meet the twin goals of productivity and participation, a large "rationalized" cooperative of production was created as the vehicle of the agrarian reform, oriented to the maximization of profits. Peasants, on the other hand, were extremely concerned with security and survival in a shifting context of uncertainty and exploitation. A strong household basis to their economy was a reflection of this concern. These tensions began to surface in the implementation of the agrarian reform. The government, responding to the dilemma, chose to stress stability and productivity, in its concern with the political trajectory of the cooperative. Participation has been the victim, though, in the process.

9

Postscript: After Five Years

With the death of Gen. Juan Velasco Alvarado on Christmas Eve of 1977, Peru lost its last remaining link with the revolution ushered in by the military coup led by Velasco in October 1968. His death touched a nerve among his loyal followers. A ground swell of emotion, including an estimated crowd of two hundred thousand that lined the street to glimpse a view of his coffin, bore testimony to his charismatic hold over the masses. The nature of his charisma will undoubtedly occupy the minds and imagination of scholars and common folk for the years to come.

The eclipse of Velasco began earlier. Following a period of ill-health and economic crisis, he was removed from office in August 1975.[1] Under his successor, Gen. Francisco Morales Bermudez, the Peruvian Revolution has changed directions. By 1977, Peru is said to have entered phase two of the revolution, marked by the "consolidation" of the reforms of the Velasco period and the shift away from a rhetoric of participation to a moderation of the social, political, and economic forces unleashed by the reforms. This shift is subject, of course, to varying interpretations. According to the government, it is a "maturation" of the revolution and a new sense of "responsibility" of its beneficiaries. More skeptical observers, particularly those associated with the purse strings of the world's capital, see it as a necessary move to cool off an expansionary and heavily debt-ridden economy in order to gain a measure of "fiscal responsibility." And leftists of various persuasions who cautiously supported the first stage of the revolution now see the government as moving to the right and reneging on the pledges to promote popular participation and structural change. Whatever one's interpretation of these events, it is

1. See David P. Werlich 1977 for an account of the events leading up to the removal of Velasco.

clear that the Peruvian Revolution is in a state of crisis. The current shift in rhetoric is one element of this crisis, the implications of which remain to be clarified.

Here we will review continuity and change in the patterns that were isolated in the earlier chapters. Our time frame will be the five-year period from 1972 to 1977. It is hoped that through such a review the implications of the shift in policy at the national level for local level implementation of the agrarian reform can be illuminated.

Agrarian Reform in the Pampa de Anta, 1972–1977

In late 1971, a crisis had beset the embryonic Túpac Amaru II cooperative. The crisis was ostensibly caused by dissatisfaction with the appointed administrator's performance and an apparent discrepancy between his salary and the incomes of the cooperative members. Underlying this surface conflict, however, were the factors that I have isolated previously in this book. At any rate, the cooperative was declared in a state of "intervention" and a commission of officials from the Ministry of Agriculture and SINAMOS arrived to rectify the situation. Their solution was to dismiss the administrator and to reorganize the cooperative along the lines described in Chapter 5.

In 1977, the cooperative was again in the midst of a crisis and in a state of intervention. Another commission had just arrived, which although composed differently with CENCIRA officials now replacing the SINAMOS contingent,[2] was following the same procedures in reviewing the state of the cooperative. As before, they perused the available records and were interviewing heads of local production units. They were still at their tasks and had not yet made their recommendations by the time I had to leave the region.

What were the elements of the crisis faced by the cooperative in 1977 and to what extent do they represent recurring

2. By 1976, in line with the shift in rhetoric, SINAMOS was removed from an active role in the agrarian reform, with its promoters disappearing from the countryside. Training functions passed over to CENCIRA and the Ministry of Agriculture assumed the role of financing and advising reform enterprises. Instead of promoting "participation" and consciousness raising, stress was now placed on management and administrative skills (compare Mariano Valderrama 1976:101–2).

leitmotifs as found in this study? First, my initial impression in traveling through the Pampa de Anta was a sense of some subtle and possibly significant changes occurring in the landscape. On the main Cuzco-Lima road that bisects the Pampa, I came across a large set of cooperative storage buildings on the entrance to the crossroads town Izcuchaca. About ten kilometers past Izcuchaca, I saw a large cattle complex complete with covered stalls and facilities for dairy operations. They appeared to be clean, well constructed, and well maintained. Farther still, on the grounds of what used to be a major hacienda, La Hoya, a set of silos had been built and classrooms outfitted for instructional purposes within the main building. Lastly, I noticed numerous Eucalyptus trees planted on many of the slopes of the Pampa that were bare and eroded in 1972. These visual impressions led me to believe that some capital investments had been made in the physical infrastructure and that they perhaps signaled important developments in economic policy and planning.

These changes were indeed closely tied to shifts in the economic policy of the cooperative during the period 1972–1977. These shifts, in turn, contributed directly to the crisis of the latter years.

Economic Policy and Performance in the Cooperative, 1972–1977

During the earlier years, cooperative agricultural strategies relied heavily on traditional crops, particularly, potatoes, barley, quinoa, and broad beans (*habas*). In 1972, as Figure 9–1 shows, over half the cultivable land was used to plant potatoes. After difficulties with potato production and marketing, as described in Chapter 6, the input of potatoes into the crop mix decreased to around 20 percent in 1975. In its place, more land was cultivated in grains, notably oats and wheat. This shift is due to a combination of several factors. First, there was an international shortfall in grain production due to the SAHEL drought, flooding in Mississippi, and poor harvests in the Soviet Union. This shortfall had the effect of raising the international price for grains. Second, because of developments in the Middle East, the cost of petroleum-derived production inputs such as insecticides and pesticides increased considerably, driving up the price of foreign agricul-

Figure 9–1. Land Use Trends in Cooperative Túpac Amaru II, in Hectares, 1972–1975

Crop	1972	1973	1974	1975
Potato	611	514	253	200
Barley	343	?	21	44
Oats	105	229	95	283
Wheat	—	669	979	299
Quinoa	27	72	39	56
Habas	28	35	27	64
Corn	1.5	15	15	56
Peas	13	?	15	?
Miscellaneous	9	15	15	9
Total	1,137.5	1,549	1,459	1,011

Source: IICA-CIRA, CENCIRA, FAO 1976, Table 7, p. 74.

tural commodities even more. Third, internal production of grain suffered because of climatic constraints and marketing obstacles. Lastly, and perhaps most importantly, the choice of grain reflected the shift to profitable crops that responded well to economies of scale and mechanization.

The availability of agricultural machinery was facilitated through long-term loans obtained through the Banco de Fomento Agropecuario, beginning in 1972. These loans have been used to obtain agricultural machinery and equipment. By 1976, the cooperative had purchased fourteen tractors, and other machinery including harrows and threshers (IICA-CIRA, CENCIRA, FAO 1976:74). Where potatoes were cultivated in ex-hacienda land adjacent to Rumipata in 1971–1972, quinoa was found in 1977.

During the earlier period, agricultural activities were given higher priority than cattle raising and dairy operations, even though significant amounts of cooperative land was suitable only for pasture. Some of this land, on which cattle was raised by the haciendas, was plowed up for potatoes. And, as we have seen, this shift was laden with obstacles. In the ensuing years since 1971, cattle raising and dairy operations have replaced agriculture, at least agriculture of the labor intensive variety, in the priorities of the cooperative. By 1975, one hundred fifty head of cattle had been purchased to increase the stock obtained initially through transfers and natural increase. Other capital investments in infrastructure and processing facilities were added. By 1975, cheese and butter, together with *chuno* and *moraya* (processed through dehydration from potatoes), brought in over 525,000 soles into cooperative coffers (IICA-CIRA, CENCIRA, FAO 1976:78–82).

Another program that has been given high priority is forestation. The Eucalyptus trees had been planted by cooperative labor in the areas that were marginal for other uses and were showing signs of erosion. At the time of my visit, these trees looked to be about five feet in height. Eucalyptus trees provide an excellent building material and cooperative officials claim that the forestry program alone will be sufficient to keep the cooperative solvent in the future.

The intensification of trends toward a "rationalization" of the cooperative through the adoption of modern technology and a concern for increased production is, in part, a reflection of the government decision in mid-1976 to de-emphasize participation and structural change and, instead, "consolidate" the advances of the revolution. At that time, there was an economic crisis, including shortages of basic foodstuffs caused by a drop in internal production and international shortfalls. Government policy in the mid-1970s was based on that crisis as well as the consistent political stance of providing food at low prices for essentially urban consumers (Harding 1975).

One could just as well question, however, whether alternative production strategies designed to utilize more labor were ever considered. For example, no attempts were made to use small-scale, more appropriate technological solutions to increase productivity that would have provided more opportunities for labor. Nor were crops chosen that might have produced better opportunities for labor use. Peasants in independent communities had found, for example, that onions grew well in the Pampa and both extended the length of the agricultural cycle and utilized more labor during the cycle.

At any rate, the problem of labor was a crucial element of the crisis faced by the cooperative. The problem was twofold. First, there was not enough labor demand generated by the cooperative's activities. One adviser told me that there was only enough work available for two hundred, while he gave me a figure of three thousand for the total membership of the cooperative. Second, in 1977, a cooperative worker was paid 120 soles a day as compared with the ongoing wage rate of 10 soles in the adjoining peasant communities. These wage levels explain the increase in fixed costs for labor from 644,983 soles in 1971 to 3,580,404 soles in 1974 (IICA-CIRA, CENCIRA, FAO 1976:82).

This situation would not be so inequitable were there comparable increases in the income of the non-beneficiary sector. Such is not the case, however. Their position is identical to that obtained in 1971–1972: There are no viable subsidized sources of credit for small holders; extension services have ignored their needs in favor of the reform sector; foodstuffs are kept low in price in order to appease the urban middle classes; and the same intermediaries exploit the weaknesses in marketing channels.

An example of the weakness of the small holder sector is the failure of the shift to onion cultivation noted in Rumipata in 1971–1972 to "takeoff." Only two plots could be found in my visit in 1977. It is difficult to reconstruct the turn of events that led to this failure due to a lack of time. I was able to elicit some reasons, however. Onion seed was said to have increased dramatically in price, making it out of the reach of any but the wealthiest peasants. Further, overproduction in the Cuzco area had resulted in low prices, and some peasants had been unable to recover costs. Some informants attributed the price fall to competition from large-scale onion plantings by the cooperative. Although onions are included under miscellaneous in the accounting system of the cooperative making it difficult to follow onion production patterns, this explanation is indicative of attitudes toward the cooperative. At the time of my visit, prices for onions were high but peasants were reluctant to plant. The few that were willing said they would seed onions in their irrigated plots after maize had been harvested, in effect voicing less than a full commitment to cash cropping.

A third element in the crisis of the cooperative was its precarious financial position. To begin, when judged against the performance of other agrarian reform units in southern Peru, the Túpac Amaru II cooperative is said to be quite well run (Gustavo Adolfo Vera 1976:91). Some evidence of its performance can be found in an IICA-CIRA, CENCIRA, FAO study, where the volume of sales is calculated to have increased from 2,039,322 soles in 1971 to 17,519,585 soles in 1974 (1976: 82). The cooperative is still subject, however, to the same exploitative marketing channels as found during the period prior to the agrarian reform. The intermediaries continue to be the major vehicles for the marketing of agricultural products. While a government agency, the Empresa Publica de Servicios Agropecuarios, was established in 1970 to intervene in the marketing of basic foodstuffs, it is involved in only a partial marketing of the potato production of the cooperative. Most products are sold to the same middlemen who purchased goods from the haciendas. They in turn sell barley to a regional brewery, wheat to regional mills, and the remainder of products to retailers in Cuzco and Abancay.

The major problem in the financial position of the cooperative, however, is its debt obligations. First, as of 1976 a total of 46,277,313 soles were owed on the land, cattle, and installations transferred to the cooperative during successive adjudications (IICA-CIRA, CENCIRA, FAO 1976:84). The terms of these transfers included a five-year grace period during which time the cooperative was responsible only for meeting interest payments of 7 percent. The first installments on reducing the principal were due in 1977, but, on my visit there early that year, nothing had yet been paid and it was not generally acknowledged by the members I consulted that they now fell due.

A second component of the debt load is the large amounts of credit obtained to finance the capital programs. Most of this credit, around 80 percent, has been short-term credit used for basic production inputs such as fertilizer and insecticide. Long-term credit, of up to twenty-years duration, has been used for the purchase of agricultural machinery and other capital investments.

The imminence and the size of debt obligations of the cooperative create a new and ominous aspect of its current crisis, with several implications. First, it points toward the need for better financial management practices including a belt tightening and the ever-present centralization of policymaking in the ranks of outside advisers and administrators. One of the most obvious corrections is likely to be a reduction in wages in order to lower the very high fixed costs for labor. This would probably be refused by the membership, although it would help to reduce the inequities in the regional distribution of income. Such a move would certainly lead to considerable antagonism toward higher level officials. Second, there will be the inevitable calls for the cancellation of the debt obligations by spokesmen for reform beneficiaries. Given the current political climate, it is unlikely that this request will be granted. Third, any hesitation, tardiness, or reduction in the debt repayment schedule is sure to give ammunition to groups opposed to the agrarian reform. Middle-class urban consumers already blame the performance of the reform enterprises for the distribution problems, and this discontent could lead to further alliances with ex-landowners. And, lastly, to meet the repayment schedule, some reallocation of funds is obvious thus creating the possibility of less cash available to meet short-term and long-term production and capital inputs. This would have the effect of lowering production and productivity, further decreasing the contribution of the cooperative to the regional economy.

Socioeconomic Differentiation

Economic problems proved to be one element of the crisis of the cooperative; a second stemmed from changes in the regional social structure associated with the agrarian reform.

To begin, the cleavages that existed in the region prior to the formation of the cooperative continue to provide bases for differentiation within the membership. These cleavages are based on different degrees and modes of relation to the ex-haciendas. For example, within the ranks of the membership there are members of recognized peasant communities, former *feudatarios,* former *feudatarios* who are also members of peasant communities, and middle holders who are usually mestizos and until recently have not entered into the affairs of the peasant communities. The interests of each of these economic groupings often are at odds, as we have seen in the preceding chapters.

One of the most significant of the preexisting groupings is that of former *feudatarios*. Although the practice is illegal, these peasants continue to control the usufruct plots they obtained under the hacienda system. Their desire to retain these plots is quite understandable given the unstable condition of the cooperative. This land could be taken away only at great cost to the government. The persistence of de facto private plots within the Peruvian agrarian reform sector is commonplace. In one study it was found that, in 1974, 40 percent of the agrarian reform cooperatives still contained private parcels held by former *feudatarios* (Cynthia McClintock 1974). In some cooperatives, the private holdings are growing rather than declining. The author found that ex-*feudatarios* take little interest in the activities of the cooperative. This comment is revealing and suggests that, as in Tukiwasi, they continue to be found in communities with a hacienda-like sociopolitical structure that isolates them from the larger society.

Of particular importance are the seasonal laborers employed by the cooperative. Because peak periods of labor demand in the cooperative often overlap with similar periods in the peasant economy, there is the need for a pool of occasional labor paid on a day basis. These peasants are not allowed to join the cooperative, however, and their status is ambiguous. Their main concern is job security, and they do not share in the interests of other members such as the demand for high-quality pasture. Although it is unclear how many individuals are employed on an occasional basis, they do not present the same problems as in the coastal cooperatives where numerically they constitute a major percentage of the work force (Alfonso Chirinos Almanza and Willy Caldas Zamudio 1976).

A second dimension of the problem is an increasing social and economic distance between cooperative members and nonmembers. In 1971–1972, the salary of the appointed administrator was so great as to create a crisis, leading to government intervention, his firing, and a reorganization. Wage differentials between cooperative members and nonmembers during the same period were on the order of two to one. By 1977, however, the ratio jumped to twelve to one. Aside from creating large increases in fixed costs for labor in the cooperative, these increases had the further effect of creating incipient class divisions. This pattern has occurred elsewhere in Peru leading to the creation of an agrarian reform elite in the rural sector.[3]

3. Ramon Zaldivar found wage and salary increases in Paramonga on the coast to range from 600 percent in the lowest class, up to 4,000 soles, to 1,400 percent in the 15,000–20,000 soles class in 1971. He argues in his analysis, however, that the raises are justified but inequitable since technicians and administrators receive the bulk of the increases (Ramon Zaldivar 1974:36–37). Comparable data are not available for the cooperative Túpac Amaru II. Colin Harding (1975)

Increasing socioeconomic differentiation between beneficiaries and non-beneficiaries and within the beneficiary class is one of the criticisms most often voiced of the agrarian reform (Conlin 1974; Harding 1975, 1977; Rubin de Celis 1977; Eguren Lopez 1975; Atusparia n.d.: 24–25). It has had the further effect in the Pampa de Anta of reinforcing the position of the intermediary groups that, while in a situation of flux during 1971–1972, have now come to play a dominant role in the new social structure.

In earlier chapters, we mentioned the dilemma represented by the *feudatarios* in hacienda communities such as Tukiwasi. The dilemma was a result of the socio-cultural isolation of the *feudatarios,* fostered and maintained by the hacienda structure. As we said, work organization and virtually every sphere of the activity of *feudatarios* were dependent on decisions made by the administrator of the hacienda. Following the agrarian reform, however, haciendas were expropriated and *feudatarios* became cooperative members. Because of their inexperience in autonomous decisionmaking, it fell upon the ex-hacienda administrators to take the lead in preparing their entrance into the larger socioeconomic-political picture then emerging. From the perspective of the reform agencies, however, hacienda administrators, and, in extension, any group dependent on the hacendado class was by definition an enemy of the agrarian reform. Phrased differently, the government was faced with a choice of keeping key local level individuals such as hacienda administrators in their positions and thus guaranteeing some degree of continuity and, most importantly, productivity, or removing them and allowing *feudatarios* to autonomously take the steps toward full participation. Our data from Tukiwasi in the period 1971–1972 indicated that there was an uneasy accommodation of MCS within the cooperative structure but there were also signals that he would eventually have to leave. By 1977, the government chose to keep MCS, giving him a position as a chauffeur for the cooperative. He continued to exert a major influence on activity in Tukiwasi and relations between local cooperative members and the co-

---

notes a general tendency among the sugar cooperatives toward the maximization of beneficiary income combined with limitations on recruitment of new members. He attributes this to group self-interest (*egoismo del grupo*). See also Peter J. Knight (1975:364–65) and E. V. K. Fitzgerald (1976:31).

operative administration. By this decision, it became apparent that MCS's key role was now institutionalized. In effect, MCS has established clientele relationships with the reform agencies that, on the one hand, guarantee his position at the local level and, on the other, provide the government with the continuity it obviously feels it needs. One can conclude that in Tukiwasi the basic hacienda structure is now firmly entrenched in the post-agrarian reform social structure.

A similar pattern has emerged in Rumipata. The schoolteacher who had established himself through astute politicking as spokesman for the community in 1971–1977 continued to occupy this crucial role in 1977. Most of his success is due to the cultivation of close personalistic relationships with the director of a key government agency in Cuzco. The latter individual represents a middle level "contact point" that, as elsewhere in Peru (Atusparia n.d.), became vulnerable to job insecurity. There are two reasons for this vulnerability. First, the increasing centralization that has occurred tends to further eclipse the input of middle level individuals by decisions made from above. Second, there is evidence of peasant oversight of the performance of middle level officials as manifested by the removal of the cooperative administrator. This oversight, however, is exercised not through "normal" channels set up by the various reform agencies, but, rather, through tried and true extra-normal channels: visits to Lima to see the president, and public meetings of peasant organizations in Cuzco calling for the replacement of incompetent and conniving officials. The latter has not effectively represented an efficient check on abuses; they continue to occur. But, as a result, they do force middle level officials to seek out support from among local level "peasant leaders" to use should any challenge against their positions emerge.

This situation has been facilitated further by an apparent reversal of a demographic trend immediately following the Agrarian Reform decree. This trend was the return to the communities of those individuals who had left earlier to live in Lima and other urban areas. They retained access to their land in the community, however, through various forms of indirect usufruct. Although I was not able to replicate the life history sample that was used in the original research to document this trend, I did attempt to ascertain whether it still held. I found that most individuals had once again returned to the urban areas they had left in the earlier period. The effect of this most recent shift was to remove from the community a group that offered significant competition for the crucial broker roles. Rumipata, for example, was not rife with the open factions that had weakened it during 1971–1972. The schoolteacher was thus able to solidify his political base both internally and externally through transactions with key agency representatives. This shift suggests, moreover, that indirect usufruct, although illegal, and an open land market, also prohibited, have reemerged in the community.

A last indicator of continuity in broker relationships comes from inquiries I made into EP's status. As we saw in an earlier chapter, EP was a "peasant leader" who manipulated real or putative relationships with peasants in the Pampa de Anta to establish himself as a radical spokesman who must be contended with. During the intervening years, he had visited Cuba and returned to the region to "instruct the campesinos." My probings among peasants revealed a new receptivity to EP which was in stark contrast with the atmosphere of antagonism toward him that I had found earlier. On the other hand, my inquiries into EP's status among government officials was answered by a guarded acknowledgment that while EP was not a formal official of the cooperative he did continue to exert considerable influence.

Size, Scale, and Structural Incongruity

The basic problems of size and scale continue to plague the cooperative. In 1973, twenty-nine additional haciendas were expropriated and incorporated into the cooperative. This action increased the land mass by 4,870 hectares and brought with it a large "captive" population of *feudatarios*. The exact size of the population of newly incorporated *feudatarios* is unclear, as is the current total membership of the cooperative. The IICA-CIRA, CENCIRA, FAO study gives a figure of 5,426 (1976:74), while government officials close to the cooperative cited a figure of 3,000. Because of the reluctance to share benefits, it would appear that cooperative members would oppose any large-scale recruiting drive aimed at the adjoining peasant communities.

The Túpac Amaru II cooperative remains a large, unwieldly, complex organization. There has been no attempt to restructure it into a smaller-scale enterprise or set of enterprises. On the contrary, the cooperative is intent on increasing the scale of operations through the choice of capital intensive, labor-saving, machinery. This results in a strengthening of the upper levels of management, both in terms of day-to-day production activities and in relation to long-term decisions concerning complicated cost-benefit analyses of capital investments. The problems that largeness of scale have introduced are evident in the following quotation taken from observations of the cooperative activities in 1974:

> The coop's planning seems chaotic, and implementation even moreso. Ideally the Production Unit Heads (with input from the coop's members in their individual Production Units) should prepare production plans for each locality. The Unit Heads and Division Chiefs and central coop officials should then integrate these Unit plans into one plan for the coop, and arrange for the provision of inputs to the Units in accordance with the plan. In practice the production plans for individual Units are hastily drawn up, and the bulk of responsibility for these seems to have fallen upon the coop's technical advisers and the Ministry. Implementation of plans is poorest in the Agricultural Division due to the complexity of providing inputs for cultivation in the various sections of the immense coop. Financial mismanagement has created liquidity problems which in turn have resulted in late deliveries of seed, fertilizer, and other inputs. Labor shortages compound these problems. (Members are seeding at the same time as the coop.) Communication and coordination problems often result in production bottlenecks—e.g., having seed but no fertilizer, or having tractors and equipment standing idle in one part of the coop while work remains to be done in the other. (Horton 1974:18.8–9)

The existence of two basic types of production units, the cooperative and the peasant community, continues into the present. As such there has been no modification of the structural incongruity between the two types that plagued the implementation of the agrarian reform. This fact is noted, for example, in the conclusions to the IICA-CIRA, CENCIRA, FAO report:

> . . . in a geographic region where two types of production units (unidades empresariales) predominate, the haciendas and communities, with peculiar relations of dependence and domination, the attempt has been made to establish an associative form of pro-

> duction that, by its very character, does not take into consideration the cultural anthropological essence of the peasant communities. As a result, while the communities have ceased to exist formally in the sense that comuneros have become members and communal land is incorporated into the cooperative, they still maintain their own dynamic (dinamica propia) that is at times dissonant with the progress of the cooperative. (IICA–CIRA, CENCIRA, FAO 1976:90)

While they give the difficulties of coordinating production schedules in each type of economy as an example of the problems encountered, it should be clear that cognitive factors associated with material processes of production underlie the dual economy.

As can be expected from the foregoing, participation by members in the affairs of the cooperative continues to be extremely poor. This generalization is supported by my own interviews with members and cooperative officials and the conclusions to the IICA-CIRA, CENCIRA, FAO study (1976: 90). Management and administration of the cooperative is generally acknowledged to be the domain of the external advisers. They formulate policy options and suggest courses of action. Their suggestions are usually supported by acquiescent upper level cooperative officials and ultimately agreed on by delegates in a general assembly. Not surprisingly, the IICA-CIRA, CENCIRA, FAO study found that power continues to reside among delegates from local production units and within the administrative and vigilance councils. Further, delegates are selected using much the same criteria as in the earlier periods. However one measures participation, whether it be attendance at local meetings of members, interest in cooperative affairs, input into short- and long-term policy and activities, information about the financial, managerial, social, and political status of the cooperative, or whatever measure, one thing is clear: In no sense is the cooperative a *cooperative* as the word is commonly understood nor is it accurately characterized by any of the terms, such as self-managed enterprise, that are commonly used in Peru to refer to these units.

While observers may argue among themselves whether the particular nature of the Peruvian experiment is that of a revolution from above (Palmer 1973) or of a "Peculiar Revolution" without either radical income redistribution or mass mobilization (Hobsbawm 1975) or of an "intermediate regime" where:

> . . . presented by the weakness of the nature upper class and its inability to perform the role of dynamic entrepreneurs on a large scale. The basic investment for economic development must therefore be carried out by the state which lends direction to the pattern of amalgamation of the interests of the lower-middle-class with state capitalism. (M. Kalecki 1972:162)

It seems clear that state control over agrarian reform enterprises is a firmly entrenched pattern of the Peruvian experiment (Atusparia n.d.:34).

Given the failure of the Túpac Amaru II cooperative to emerge as a means through which peasants can participate in key decisions affecting their lives, one can ask whether the other vehicles of participation created by the government have been effective. There are two other modes of political participation charged with representing peasant interests. They are government-sponsored peasant organizations and officially recognized peasant communities.

In October 1973, SINAMOS organized the Federacion Agraria Túpac Amaru II as a vehicle to combine all of the various interest groups operating in the Department of Cuzco: the SAIS units, agrarian cooperatives, *feudatarios*, and small holders. A congress was held and representatives were named. This officially recognized organization contrasts sharply with the existing peasant organizations dating to the peasant mobilization of the sixties. Although neither is an adequate vehicle for representing peasant interests, government-sponsored efforts, as elsewhere in Peru (Susan C. Bourque and David Scott Palmer 1975) have been marked by containment, cooptation, underfunding, and attempts to undermine leadership.

The Peasant Communities Statute of 1970 is an important cornerstone of government rural development policy. It was designed primarily to facilitate collective forms of resource use, notably through the restructuring of land tenure, and to set up political linkages with key government agencies, in particular the Peasant Communities Agency. In our research during the 1971–1972, we found that the major change brought about by the statute was the holding of new elections that caused a power struggle in each community. This unintended effect of the statute has occurred throughout Peru wherever we have adequate documentation (Norman Long and David Winder 1975; David Scott Palmer and Kevin Jaj Middlebrook 1976). Once these battles were fought in the early 1970s, each community readjusted to the new political structure. Aside from these new factors at the local level, the Peasant Communities Statute has introduced no new patterns of participation. Instead of providing a vehicle for ongoing and viable linkages between peasant communities and key government agencies, it has institutionalized the old game of political entrepreneurs who deliver support and peasant acquiescence to the agencies in return for favors, both material and nonmaterial. Land tenure patterns, in particular, the pattern of small individual holdings on the bottomlands that contrast with communal control and individual usufruct plots on the slopes of the mountains, have not been restructured, which is probably for the better. In the larger picture of government rural development policy, peasant communities are actually at a disadvantage, since agrarian reform enterprises are given preferential access to credit, technical assistance, and other services. The contrast between the highly visible, capital intensive agrarian reform enterprises and the undercapitalized impoverished peasant communities is glaring.

Summary and Discussion

A number of factors that militated against the full participation of beneficiaries in the Túpac Amaru II cooperative were isolated in my original research in 1971–1972. I also expressed the hope that adjustments would be made to resolve these contradictions. My review of the development of the cooperative in the period since then reveals an amazing degree of continuity. The same obstacles to participation and economic performance are still found. They include the questions of size and scale of the cooperative as production unit and the underlying structural incongruity between it and the independent peasant economy. With the exception of some increase of size brought about by the incorporation of expropriated haciendas, government policy has not changed these basic parameters. Indeed, the questions of size and scale do not seem to be understood at all. If anything, solutions to the crisis of the Túpac Amaru II cooperative, notably its transformation into a SAIS (IICA-CIRA, CENCIRA, FAO 1976: 91), or, as suggested elsewhere, the incorporation of agrarian reform cooperatives into even larger agricultural units or the new social property sector (Knight 1975:364–65), reflect a tendency toward larger scale.

Nevertheless, there have been some significant shifts in the five-year period. These include a worsening of the cooperative's economic position and increasing socioeconomic differentiation within the membership and between it and the nonbeneficiary peasant sector. While "efficiency" may or may not be a valid measure of economic performance,[4] it does appear that the goals set for the cooperative by the government are along the lines of increases in production and productivity rather than in the provision of opportunities for labor absorption. This can be the only explanation for the shift to a capital intensive production strategy, with a concomitant decrease of labor in a labor surplus economy. The labor problem is thus one element of the current economic crisis of the cooperative. Others include the large increases in wages and salaries that reduce the surplus available for capital accumulation and the imminence of the end of the grace period and the assumption of an onerous debt repayment schedule.

The lack of labor opportunities has to some degree offset the tendencies toward the creation of an elite class of beneficiaries created by the salary and wage increases. Nevertheless, these tendencies exist. Unlike other areas of Peru (compare

4. The question of the "efficiency" of the Peruvian enterprises is often cast as if it were necessary to set them in contrast with private enterprise. One must be careful here. As Fitzgerald points out (1976:54), there are a number of special features of the Peruvian situation that make this comparison meaningless. First, private enterprises in Peru has not been the engine of development the effects of which are felt throughout all sectors of the society. Second, although foreign capital could have possibly undertaken these tasks, the explicit aim of the state has been to reduce this kind of external dependency. Third, the goals of the reform enterprises should not necessarily be profit maximization as such but rather contributing to the economic goals of the state, which may, of course, include the creation of labor opportunities. While "efficiency" may be one criterion to evaluate economic performance, the basic point is that these criteria are not necessarily the same as those found in private enterprise.

Harding 1975), there has been a remarkable calm in the Pampa de Anta. The socioeconomic differentiation introduced by the agrarian reform has not produced any overt manifestations of discontent, except for some isolated invasions of cooperative land by neighboring peasant communities in the early 1970s. In part, a limited need for temporary laborers and the difficulties of communication accounts for the pattern. I would suggest, however, that a major reason is the creation and manipulation of clientele relationships established by key members of government agencies with local political entrepreneurs in each community that has allowed government agencies to co-opt and contain local level tension and conflict.

In light of the foregoing, the current "crisis" of the cooperative has a strong sense of déjà vu. Based on our knowledge of the 1971–1972 situation, one would have predicted the occurrence of such crisis unless the necessary structural changes in the cooperative were made. The only difference between the crisis and mode of response is that the impending schedule of payments on cooperative debt makes the current crisis more profound.

The amazing continuity manifested in the patterns of participation in the Túpac Amaru II cooperative sheds light on the larger patterns of the Peruvian agrarian reform. It suggests that the shift in government rhetoric from a stance of "participation" (1970–1976) to "consolidation" (1976 to present) is illusory. Of course, one must qualify this statement with respect to the different types of agrarian reform enterprises and the region of the country. Thus, those policy shifts in the direction of participation have occurred in regions where the economy is directly tied into the export sector (Harding 1975). And, it is true, as predicted by several observers (Fitzgerald 1976:31), the land available for expropriation has been exhausted. It is most likely that the term *consolidation* best translated into a reluctance to expropriate holdings in the middle range of land tenure, and a conservative rhetoric designed to please international lending agencies. At any rate, the evidence shows that the Túpac Amaru II cooperative has never been run with participation in mind, rather it has been oriented to serving the needs of urban consumers and voters, as perceived by a centralized military government.

Appendix

Methods of Data Collection

In this appendix, I will briefly describe the types and manner in which data were collected, beginning with fieldwork in the three communities, and then discuss research strategies at the regional and national levels.

Data Collection: The Communities

My research plan involved selecting three basic types of communities in the Pampa de Anta: (1) a rich, free community with relatively few or no ties to neighboring haciendas; (2) a poor hacienda community in which most or all of the peasants had contractual ties to a hacienda; and (3) a free community that had some land but also a significant number of peasants also working on an adjacent hacienda. Hopefully, at least two of the communities would be adjacent so that I might work in both at the same time.

After examining the characteristics of the communities that were available to me in the various censuses, I made several trips out to the Pampa to look over possible sites. I made one trip with Dr. Demetrio Roca Walparimachi of the Anthropology Department of the University of San Antonio Abad in Cuzco. Dr. Roca is a native Anteño and has considerable knowledge of Pampa communities. With his advice and a tentative list of communities drawn from the censuses, I finally selected two communities: Rumipata, which is one of the largest and richest communities in the Pampa in terms of land, and Tukiwasi, a rather small hacienda community where plots set aside for peasants average about one *topo* in size. These communities were adjacent and afforded me the possibility of working in both at the same time. A third community, Antapampa, was picked after about six months of fieldwork.

Fieldwork in these communities was carried out in residence between November 1971 and August 1972. During this period, my wife and I rented homes in the communities, participated in the lives of the people living there, and informally interviewed residents. These were our major data-gathering techniques; we supplemented them on occasion with more formal survey techniques.

Fortunately, prior to our arrival a census of the communities had been made as a step on the implementation of the agrarian reform. This census, which I have referred to as the Peasant Communities Census, was carried out during May to June of 1970. Teams of trained census takers were sent to the region, first to notify and inform the communities of the upcoming census, and then to return a week later to enumerate the assembled residents. During the next few days, the teams then went from door to door to locate any peasants who had been missed for one reason or another in the first census. Throughout the process, it was stressed that this was an official census and that each household head was legally responsible for its contents. The census form itself was rather long (eight pages) and detailed, containing questions on household composition, occupation, agricultural production, and a series of questions on the amount, kind, and quality of land held by each household.

Results of the census have been reported in two volumes. The first (Ministerio de Trabajo 1970) provides a summary of demographic and economic indices for the region. There the census universe is analyzed in terms of populated centers, *centros poblados*, following the usage of the 1961 national census. In the introduction to the first volume, the populated centers are divided into four categories. Recognized peasant communities are those that are recognized under the provisions of the 1920 constitution; unrecognized peasant communities lack such recognition. Annexes are separate autonomous peasant communities that because of historical accident have separated from established communities, yet continued to maintain representation in the larger community. The *fundo/hacienda* category, which is referred to as hacienda in this study, is generically different from the peasant community. In the Peasant Communities Census, only those haciendas with resident peasants, or *feudatarios*, were censused. In the analysis of census data from haciendas, while seventy haciendas were censused, only forty were eventually tabulated; in the other thirty haciendas, peasants asked to be enumerated in the populations of adjoining peasant communities that they claimed as their residence. Figure 10–1 summarizes the distribution of types of populated centers in the Peasant Communities Census.

Figure 10–1. Distribution of Communities in Universe: Peasant Communities Census

District	Recognized Community	Unrecognized Community	Annex	Hacienda	Total
Anta	6	4	5	13	28
Zurite	11	2	1	15	29
Huarocondo	6	1	1	12	20
Total	23	7	7	40	77

Source: Derived from Table 1 in Ministerio de Trabajo 1970:6.

In the analysis of the Peasant Communities Census, in both volumes, recognized, unrecognized, and annexes of communities are subsumed under the rubric community; the other category of analysis is the hacienda. This is a reflection of the relative unimportance of the political criteria distinguishing the three types of peasant communities; each is essentially autonomous and stands alone as an independent economic unit.

The second volume (Ministerio de Trabajo 1971) gives a multivariable analysis of the data reported in the first volume using labor force, family income, and the total of land. Both of these volumes have been useful for the discussion in Chapters 2 and 5 of demographic and economic patterns at the regional level. With the assistance of a research assistant, I was able to retrieve data at the household level from the original census forms for the three communities then kept in the Peasant Communities Agency office in Cuzco; this gave me relatively complete census information for the households of my study communities.

Most anthropologists are not so fortunate as to be provided with such fresh quantitative data with so little effort on their part. I can certainly concur in this regard. Nevertheless the skeptic is entitled to ask the question that must come up at some point: "Well, how good is this gold mine? Aren't peasants suspicious, hostile, and certainly in the Peruvian case, exploited by people such as census takers?" As I have used the peasant communities census freely throughout the study I must address this question with some forthrightness.

One insight into the quality of the census is given by those individuals who were involved in its planning, implementation, and subsequent analysis. In the introduction to the first volume, the authors claim that the data are "80 percent approximate." This claim is based on what they feel was a social climate conducive to the administration of the census. In support of this claim, they point to the fact that peasants did collaborate in the taking of the census, there was in general a positive attitude toward the census exhibited by the peasants, and it was made clear to the peasants that the census was a legal document. Where deficiencies do occur, they are implied to stem from the low levels of literacy, education, and bilingualism.

I believe that this evaluation of the Peasant Communities Census is accurate with some exceptions. In retrospect, there was a favorable social climate among the peasant population of the region. They had heard of the decreeing of the Agrarian Reform Law and the government's application of the law in the highly productive coastal plantations and were optimistic, if not confused, as to how the agrarian reform would be implemented. The census takers could argue that it was in the interest of the peasant to respond to the questions that were asked of him; if one could establish a claim to land and to being an acknowledged member of a peasant community, then the benefits of the agrarian reform would be opened up to him. The census was perceived in this regard as an essential step in gaining benefits that were rumored at the time to include the distribution of parcels to individual peasants.

It cannot be denied, though, that there were instances of both conscious and unconscious misreporting and underreporting. In some cases, the richer peasants underreported their totals for the amount of land in their possession because of the fear that excess land would be expropriated. The same fear extended to land that was cultivated indirectly, that is, in one of the various forms of indirect exploitation explored in detail in Chapter 4. Thus, the category of land held indirectly in the Peasant Communities Census is lower than it should be. And another source of error results from the common problems of censusing what is essentially an illiterate indigenous population unaccustomed to responding to modern survey investigation techniques. Age, for example, is consistently misreported among the older segment of each community's population. In some instances, I have supplemented with the

1961 Peruvian census of populated centers (Dirección Nacional de Estadística y Censos 1974). In other cases, I gathered additional data in the field in order to extend my investigation into areas not covered in either of the censuses.

As an example of the latter, about the middle of my field period in Rumipata I became aware that some households were beginning to cultivate onions for market in place of the traditional mix of subsistence crops. This transition was important in terms of the various kinds of behavior I was investigating. Since the Peasant Communities Census did not reveal this dimension, I decided to gather data that would reveal the process in its quantitative dimension. I did this through an "on the ground" census of all the irrigated plots in the community, giving owner, type of tenure, and the size. This census was made using assistants hired from Rumipatiños who had lived in the community over a period of time. I then enumerated separately the irrigated plots in onion cultivation. This census is used in the construction of two figures in Chapter 3.

To save time in the field and to extend the depth of my study, I made considerable use of research assistants. The first assistant, an anthropology student from the University of Cuzco, had considerable success in collecting data in the offices of the cooperative in Sullupucyo and the peasant communities office in Cuzco. In June, she made a cultural survey of Antapampa while I finished my research in Rumipata and Tukiwasi. The second assistant, a graduate anthropologist from La Católica University in Lima, had been working in Antapampa on her own project involving leadership. I employed her for a month in June to continue her research on leadership and to begin informal interviewing on attitudes and behavior of the Antapampeños toward the cooperative. The work these assistants did in Antapampa was of considerable benefit to me when I finally arrived there in July.

I contracted a third assistant, an anthropology student from the University of Cuzco, to take a sample of migration histories. I was interested in migration as a factor of social change in the region and in particular the phenomenon of return migration to peasant communities following the agrarian reform. Return migrants were influential in acting as brokers in providing information to peasant beneficiaries. The student I employed had been working for Ron Skeldon, a cultural geographer from the University of Toronto, who was carrying out a study of migration in several departments of the southern sierra. I modified a structured interview schedule used in Skeldon's fieldwork and applied it in the three communities using my assistant. Interviewing was done in Quechua except when the respondent was able to speak Spanish. The sample is an availability sample; my assistant was instructed to go out and interview a certain number of adult peasants in each community. Fifty peasants were interviewed in Rumipata, fifteen in Tukiwasi, and thirty in Antapampa. In the tables in the study that report the analysis of the migration histories, the sample size for each community is smaller. This is a reflection of the quality of the interviews; some interviews did not include certain information and thus were not included in the table. Because of the nature of the sample, I have not attempted to generalize about the process of migration in the region. The sample is useful in providing insight into the migration process in the three communities and in giving depth to what might be otherwise loose observations.

The Region

The Peasant Communities Census was an important resource in my next task: the description and analysis of the social and physical environment of the Pampa de Anta both before and after the implementation of the agrarian reform. These subjects form the content of Chapters 2 and 5.

Data at the regional level came from a number of sources. In Lima, I obtained aerial photographs of the Pampa de Anta from the Military Geographic Institute. These photographs have been extremely useful in the beginning of my fieldwork in the communities, by giving one aspect of the ecology as seen from the air, and, in conjunction with maps, to lay out the settlement pattern with reference to the distribution of haciendas and independent communities. As excellent map (1:50,000) of the Pampa containing the boundaries of all the communities, towns, and haciendas was obtained from the Ministry of Agriculture in Cuzco.

I took every opportunity to visit other parts of the Pampa. On one occasion, my wife and I visited Zurite during the inauguration of an annual agricultural fair that was organized by the Ministry of Agriculture. The visit was highly instructive in revealing the dilemma of district capitals such as Zurite in maintaining a viable economic base. Much more successful was the thriving new market in nearby Ancawasi created entirely by *cholo* entrepreneurial middlemen. On other occasions, trips to and from Cuzco for provisioning brought us through Izcuchaca, the satellite town of Anta, the provincial capital. Izcuchaca is the transportation and communications hub of the region and is an important locus for mestizo social interaction.

Data concerning the implementation of the agrarian reform have come from a number of sources. One source has been the collection of newspaper articles on the agrarian reform and related events. This task was done with the aid of a research assistant and included all articles since the enactment of the law from *El Comercio* of Cuzco.

Another important source of information was the files of the Ministry of Agriculture in Cuzco. I worked during the first few months upon arriving in Cuzco in the Ministry of Agriculture and was able to fill in the background of the creation of the bureaucracy described in Chapter 5.

A final source of information on the agrarian reform at the regional level is a set of reports made as the result of fieldwork by anthropology students from the University of Cuzco. During 1970, students were sent to individual communities in the Pampa to research a set of topics that had to do with its imminent implementation. This work was done in conjunction with the Agrarian Reform Agency of the Ministry of Agriculture and contains considerable data on attitudes of peasants toward the agrarian reform before it became a reality.

The quality of these reports varies considerably; while some are virtually useless, others approach the standards of professional ethnography. Taken together, they did give a general insight into the confusion in the minds of the peasants as they contemplated the then imminent expropriation and distribution of land.

Students were sent into the region again in 1972 to evaluate the implementation of the agrarian reform, in particular, the Túpac Amaru II cooperative. A number of problems plagued their efforts and I had to leave the country before their reports that had been delayed were written up. In many respects, their problems were illustrative of the problems I was then experiencing in my own work. They involved the reluctance on the part of technicians and advisers associated with the cooperative to cooperate with social scientists whether they be Peruvian students or foreign academics. This unwillingness emerged in middle-late 1971 and became particularly acute following the failure of the Alto Urubamba cooperative in the Convención Valley in February 1972 which received considerable press attention. In essence, these individuals represented the interests of both the cooperative and the Ministry of Agriculture; in their latter capacity, they were especially fearful of criticism of their efforts during a crucial transition period to a new agrarian structure. As discussed in Chapter 4, such an overly cautious attitude has been characteristic of much of the behavior of governmental agencies, particularly in the Pampa de Anta, which was a "showcase" (John Gitlitz 1971) for the agrarian reform in the southern sierra.

The Nation

A third type of data is that bearing on the role of the Agrarian Reform Law as a national policy, that is, its history, the political context in which it emerged, and its relation to other reform measures passed by the Velasco government.

Part of this research task was facilitated by a mandatory trip every three months to Lima to renew my visa. These trips enabled me to survey the bibliographic literature of the agrarian reform at the level of the government ministries in Lima, including the Ministry of Agriculture, the Peasant Communities Section of the Ministry of Labor, and the Agrarian Reform Agency of the Ministry of Agriculture. I established early contact with the Center for Research and Development in the Agrarian Reform, and after giving a lecture on my proposed research was given access to their library that contains virtually all the Peruvian and foreign literature to come out on the 1969 Agrarian Reform Law. The bibliographic literature on the 1969 Agrarian Reform was surprisingly limited at that time, and to supplement it I had to resort to informal interviewing both in Cuzco and Lima. In Lima, I was fortunate enough to stay in the home of Juan Languasco, the ex-minister of government under President Belaúnde and the owner of a large cotton plantation in Piura Department that was expropriated by the agrarian reform. Through him and other individuals, I obtained the landowner's perspective of the political context in which the Agrarian Reform Law emerged.

Another source of data that had earlier proved useful was the newspaper archives. Again, with the aid of a research assistant I collected all of the articles on the agrarian reform taken from *El Peruano,* the official government paper, and *La Prensa,* a conservative Lima daily.

Bibliography

Acheson, James M.

1972*a* "Accounting Concepts and Economic Organization in a Tarascan Village: Emic and Etic Views." *Human Organization* (Spring) 31:1:83–91.

1972*b* "Limited Good or Limited Goods? Response to Economic Opportunity in a Tarascan *pueblo*." *American Anthropologist* 74:5:1152–69.

Adams, Richard N.

1962 "The Community in Latin America: A Changing Myth." *The Centennial Review* 6:3:409–34.

1970 "Brokers and Career Mobility Systems in the Structure of Complex Societies." *Southwestern Journal of Anthropology* 26:4: 315–27.

Adolfo Vera, Gustavo

1976 *Cuzco: reforma agraria y cambios en la propiedad de la tierra 1969–1974*. Lima: Universidad Mayor de San Marcos. Seminaria de Historia Rural Andina.

Alers, J. Oscar, and Appelbaum, Richard P.

1968 *La migracion en el Peru: un inventorio de proposiciones*. Lima: Centro de Estudios de Población y Desarrollo.

Arnold, Dean

1975 "Ceramic Ecology of the Ayacucho Basin, Peru: Implications for Prehistory." *Current Anthropology* 16:183–205.

Atusparia, Pedro

N.D. (1976?) *La Izquierda y la Reforma Agraria Peruana. Tres Cuestiones Fundamentales*. Lima: Editorial Difusión Popular.

Austin, Allen

1964 *Estudio sobre el Gobierno Municipal del Perú*. Lima: Oficina Nacional de Racionalización y Capacitación de la Administración Pública.

Bailey, F. G.

1970 *Stratagem and Spoils*. New York: Schocken Books, Inc.

Barth, Fredrik, ed.

1963 *The Role of the Entrepreneur in Social Change in Northern Norway*. Oslo: Norwegian University Press.

Bennett, John W.

1966 "Further Remarks on Foster's 'Image of Limited Good.' " *American Anthropologist* 68:206–10.

Bolton, Ralph, and Mayer, Enrique, eds.
1977 *Andean Kinship and Marriage.* Special Publication No. 7. Washington: American Anthropological Association.

Bonilla, Heraclio
1967–1968 "La Coyuntura Comercial del Siglo XIX en el Perú." *Revista del Museo Nacional* 35:159–87.
1974 *Guano y Burguesía en el Peru. Peru Problema 11.* Lima: Instituto de Estudios Peruanos.

Bourricaud, François, et al.
1971 *La Oligarquía en el Perú.* 2d ed. Lima: Instituto de Estudios Peruanos.

Bourque, Susan C.
1971 *Cholificatión and the Campesino: A Study of Three Peruvian Peasant Organizations in the Process of Societal Change.* Latin American Studies Program Dissertation Series No. 21. Ithaca: Cornell University Press.

Bourque, Susan C., and Palmer, David Scott
1975 "Transforming the Rural Sector: Government Policy and Peasant Response." In *The Peruvian Experiment: Continuity and Change under Military Rule,* edited by Abraham F. Lowenthal, pp. 179–219. New Jersey: Princeton University Press.

Brush, Stephen B.
1977 *Mountain, Field, and Family.* University of Pennsylvania Press.

Cancian, Frank
1965 *Economics and Prestige in a Maya Community.* California: Stanford University Press.

Castro Pozo, Hildebrando
1924 *Nuestra Comunidad Indígena.* Lima: El Lucero.

CENCIRA (Centro Nacional de Capacitación e Investigación para la Reforma Agraria)
1973 *El Flújo de las Comunicaciones en las Empresas Campesinas.* Lima.

Chaplin, David
1967 *The Peruvian Industrial Labor Force.* New Jersey: Princeton University Press.

Cheung, S. N. S.
1969 *The Theory of Share Tenancy.* Chicago: University of Chicago Press.

Chirinos Almanza, Alfonso and Caldas Zamudio, Willy
1976 *Percepción del Campesinado y los Cambios en el Sector Rural del Perú.* Lima. Universidad Nacional Agraria la Molina. Centro de Investigaciones Socio-Económicas.

Cotler, Julio
1970 "The Mechanics of Internal Domination and Social Change in Peru." In *Masses in Latin America,* edited by Irving L. Horowitz, pp. 407–44. New York: Oxford University Press.

COMACRA

1970 Modalidad de Adjudicacion del Proyecto "Anta." Lima.

Comité de Educación

N.D. *Cartilla No. 1 de Educación Cooperativa.* Cooperativa Agraria de Produción Túpac Amaru II de Anta Pampa Ltda.

Conlin, Sean

1974 "Participation Versus Expertise." *International Journal of Comparative Sociology* 15:3–4:151–66.

Craig, Wesley W.

1968 *El Movimiento Campesino en La Convención, Perú.* Lima: Instituto de Estudios Peruanos.

Dandekar, V. M.

1962 "Economic Theory and Agrarian Reform." *Oxford Economic Papers* 14:69–80.

Dirección Nacional de Estadística

1944 *Censo Nacional de Poblacion y Ocupacion de 1940.* Lima.

1966 *Centros Poblados: Cajamarca, Callao, Cuzco, Huancavelica, Huanuco.* Tomo II. Lima.

1966 *Centros Poblados: Cajamarca, Callao, Cuzco, Huancavelica. Huanuco.* Tomo I. Lima.

1969 *Anuario Estadistico del Peru 1966.* Vol. 27. Lima: Imprenta del Ministerio de Hacienda y Comercio.

Dore, Ronald F.

1971 "Modern Cooperatives in Traditional Communities." In *Two Blades of Grass: Rural Cooperatives in Agricultural Modernization,* edited by Peter Worsley, pp. 43–65. Manchester, Eng.: Manchester University Press.

Eguren López, Fernando

1977 "Política Agraria y Estructura Agraria." In *Estado y Política Agraria,* pp. 217–55. Lima: Centro de Estudios y Promoción del Desarrollo.

Einaudi, Luigi R.

1969 *The Peruvian Military: A Summary Political Analysis.* California: Santa Monica, The Rand Corporation.

Erasmus, Charles

1956 "The Occurrence and Disappearance of Reciprocal Farm Labor in Latin America." *Southwestern Journal of Anthropology* 12:444–69.

1961 *Man Takes Control.* New York: Bobbs-Merrill Company, Inc.

Fals, Borda, Orlando

1961 *Campesinos de los Andes. Estudio Sociológico de Saucio.* Bogotá: Editorial Iqueima.

Fitchen, Janet M.

1961 "Peasantry as a Social Type." In *Proceedings of the Annual Spring Meeting of the American Ethnological Society,* pp. 114–19. Seattle: University of Washington Press.

Fitzgerald, E. V. K.
1976 *The State and Economic Development: Peru Since 1968.* University of Cambridge, Department of Applied Economics. Occasional Paper No. 49. Cambridge University Press.

Flores, Edmundo, and Eckstein, Solomon
1970 "La Reforma Agraria del Perú." *El Trimestre Económico* 38: 3:515–23.

Flores, Marin, José A.
1977 *La Explotación del Caucho en el Perú.* Lima: Seminario de Historia Rural. Universidad Nacional Mayor de San Marcos.

Flores, Ochoa, Jorge A.
1970 "Algunas de las Otras Razones por las Cuales la Gente Emigra." *Wayka* 3:72–81 (Cuzco).

Foster, George
1961 "The Dyadic Contract: A Model for Social Structure of a Mexican Peasant Village." *American Anthropologist* 63:1173–92.
1965 "Peasant Society and the Image of the Limited Good." *American Anthropologist* 67:293–315.

Fuenzalida, Fernando
1969 *La Matriz Colonial de la Comunidad de Indígenas Peruana: Una Hipotesis de Trabajo.* Lima: Instituto de Estudios Peruanos.

1970 "Poder, raza, y etnía en el Perú contemporáneo." In *El Indio y el Poder en el Perú.* Lima: Instituto de Estudios Peruanos.

Gade, Daniel
1975 *Plants, Man, and the Land in the Vilcanota Valley of Peru.* Biogeographica Vol. 6. Junk: The Hague.

Gall, Norman
1971 "The Master is Dead," *Dissent* (June):281–320.

García, Antonio
1971 *Perú: Una Reforma Agraria Radical.* 2d ed. Cuadernillos de Divulgación Serie G. Artículos No. 4. Lima: Instituto Nacional de Planificación.

Georgescu-Roegen, Nicholas
1969 "Institutional Aspects of Peasant Communities: An Analytical View." In *Subsistence Agriculture and Economic Development,* edited by Clifton Wharton. Chicago: Aldine Publishing Co.

Gitlitz, John
1971 "Impressions of the Peruvian Agrarian Reform." *Journal of Interamerican Studies and World Affairs* (July-October) 13: 3:456–74.

Greenhill, Robert G., and Miller, Rory M.
1973 "The Peruvian Government and the Nitrate Trade, 1873–1879." *Journal of Latin American Studies* 5:1:107–31.

Gregory, James R.
1975 "Image of Limited Good, or Expectation of Reciprocity?" *Current Anthropology* 16:1:73–92.

Guillet, David
1974 "Transformación Ritual y Cambio Socio-Político." *Allpanchis* 6:143–60 (Cuzco).
1976 "Migration, Agrarian Reform, and Structural Change in Rural Peru." *Human Organization* 35:3:295–302.

Guillet, David, and Schaedel, Richard P.
1973 "Latin American Peasant Economies and Politics." *Journal of Interamerican Studies and World Affairs* 15:4:494–99.

Guillet, David, and Whiteford, Scott
1974 "A Comparative View of the Role of the Fiesta Complex in Migrant Adaptation." *Urban Anthropology* 3:2:222–42.

Haney, Emile B.
1969 *The Economic Organization of Minifundia in a Highland Community of Colombia.* Ph.D. diss., University of Wisconsin.

Harding, Colin
1974 *Agrarian Reform and Agrarian Struggles in Peru.* Working Paper No. 15. Center of Latin American Studies. University of Cambridge.
1975 "Land Reform and Social Conflict in Peru." In *The Peruvian Experiment: Continuity and Change under Military Rule,* edited by Abraham F. Lowenthal, pp. 220–53. New Jersey: Princeton University Press.
1977 "Agrarian Reform and Agrarian Struggles in Peru." In *Social and Economic Change in Modern Peru,* edited by Rory Miller, Clifford T. Smith, and John Fisher, pp. 120–35. Monograph Series No. 6. Centre for Latin American Studies. The University of Liverpool.

Havens, Eugene A., L., Montero, Eduardo, and Romieux, Michel
1965 *Cerete: un área de latifundio.* Facultad de Sociología. Informe Técnica No. 3. Bogota: Universidad Nacional de Colombia.

Hobsbawm, E. J.
1971 "Peru: the 'Peculiar' Revolution." New York Review of Books. 16 December 1971. pp. 33–34ff.

Horton, Douglas E.
1973 *Haciendas and Cooperatives: A Preliminary Study of Latifundist Agriculture and Agrarian Reform in Northern Peru.* Research Paper No. 53. Madison: Land Tenure Center.
1974 *Land Reform and Reform Enterprises in Peru.* 2 Vols. Report submitted to the Land Tenure Center and the International Bank for Reconstruction and Development. Madison.

IICA-CIRA (Instituto Interamericano de Ciencias Agrícolas), CENCIRA (Centro Nacional de Capacitación e Investigacion para la Reforma Agraria), and (FAO) Organización de las Naciones Unidas para la Agricultura y la Alimentación.

1976 *Cambios Contemporáneos en la Estructura Agraria Peruana.* Serie: Informe de Conferencias, cursos, y reuniones 19. Bogotá: Instituto Interamericano de Ciencias Agrícolas.

Isbell, Billie Jean
1971 "No servimos mas. . . ." *Revista del Museo Nacional* 37:285–98.

Johnson, Allen W.
1971 *Sharecropping of the Sertao: Economics and Dependence on a Brazilian Plantation.* California: Stanford University Press.

Kalecki, M.
1972 *Essays on the Economic Growth of the Socialist and the Mixed Economy.* Cambridge University Press.

Kaplan, David, and Saler, Benson
1966 "Foster's 'Image of the Limited Good': An Example of Anthropological Explanation." *American Anthropologist* 68:202–6.

Kawata, Carlos y Pierre Lobstein
1972 *El aprovechamiento de los Recursos Humanos en la Cooperativa Agraria de Producción Túpac Amaru II.* Escuela Superior de Administración Pública. Lima: Programa Experimental de Formación de Especialistas en Planificación de Recursos Humanos.

Knight, Peter J.
1975 "New Forms of Economic Organization in Peru: Toward Worker's Self Management." In *The Peruvian Experiment: Continuity and Change under Military Rule,* edited by Abraham F. Lowenthal, pp. 350–401. New Jersey: Princeton University Press.

Kubler, George
1946 "The Quechua in the Colonial World." In *Handbook of South American Indians,* edited by Julian Steward, pp. 331–410. Vol. 2.

Kuitenbrouwer, Joost
1973 *The Function of Social Mobilization in the Process Towards a New Society in Peru.* Occasional Paper No. 36. The Hague: Institute of Social Studies.

Lambert, Bernd
1977 "Bilaterality in the Andes." In *Andean Kinship and Marriage,* edited by Ralph Bolton and Enrique Mayer, pp. 1–27. Special Publication No. 7. Washington: American Anthropological Association.

Leeds, Anthony
1973 "Locality Power in Relation to Supralocal Power Institutions." In *Urban Anthropology,* edited by Aidan Southall. New York and London: Oxford University Press.

Lewis, Oscar
1951 *Life in a Mexican Village: Tepoztlan Restudied.* Urbana: University of Illinois Press.

Locker, Michael
1969*a* "Perspective on the Peruvian Military, Part 1." *North American Congress on Latin America (NACLA)* 3:5:1–10.
1969*b* "*Perspective on the Peruvian Military*, Part 2." *NACLA* 3:6:1–15.

Lockhart, James
1972 *The Men of Cajamarca. A Social and Biographical Study of the First Conquerors of Peru.* Austin: University of Texas Press.

Long, Norman, and Winder, David
1975 "From Peasant Community to Production Co-operative: An Analysis of Recent Government Policy in Peru." *Journal of Development Studies* 12:1:75–94.

Luce, Duncan R., and Raiffa, Howard
1957 *Games and Decisions.* New York: John Wiley and Sons, Inc.

McClintock, Cynthia
1974 *The Impact of Agrarian Reform Organization on the Attitudes and Behavior of Organization Members in Peru.* Unpublished Manuscript, Department of Political Science, Massachusetts Institute of Technology. (Cited in Harding 1975:224, Footnote 13.)

Macera, Pablo
1975 *Historia Económica Peruana.* Lima: Biblioteca Andina.

Mac-Lean y Estenos, Roberto
1965 *La Reforma Agraria en el Perú.* Mexico City D.F.: Instituto de Investigaciones Sociales, Universidad Nacional.

Mariátegui, Jośe Carlos
1967 *7 ensayos de Interpretación de la Realidad Peruana.* Lima: Empresa Editorial Amauta.

Medina, Rubens
1970 *Agrarian Reform Legislation in Peru.* Madison: Land Tenure Center.

Ministerio de Agricultura
1964 *Informe del Estudio Socio-Económico de las Comunidades y de los Fundos de la Provincia de Anta—Cuzco.* Ministerio de Agricultura, Files, Cuzco.

Dirección de Comunidades Campesinas
1970 *Proyecto Anta: Datos para Adjudicación.* Lima.

Ministerio de Trabajo
1971 *Cooperativa 'Túpac Amaru II' Antapampa: Datos para el Desarrollo.* Lima.

Montoya, Rodrigo, Amelia Guido, et al.
1974 *La SAIS Cahuide y sus Contradicciones.* Lima: Universidad Nacional Mayor de San Marcos.

Murra, John V.
1972 "El 'Control Vertical' de un Máximo de Prisos Ecológicos en la

Economía de las Sociedades Andinas." In *Visita de la provincia de León de Huanuco (1952). Iñigo Ortiz de Zúñiga, visitador.* Book 2, pp. 429–76. Huanuco, Perú: Universidad Nacional Hermilio Valdizan.

Na.
1971 *Plan de Capacitacion para la Poblacion de Antapampa Elaborado por sus Dirigentes.* Annex of Kawata and Lobstein 1972.

Neurath, Paul M.
1962 "Radio Farm Forum as a Tool of Change in Indian Villages." *Economic Development and Cultural Change* 10:1:257–83.

Nuñez del Prado, Oscar
1973 *Kuyo Chico: Applied Anthropology in an Indian Community.* Chicago: University of Chicago Press.

Orlove, Benjamin S.
1977 "Integration Through Production: The Use of Zonation in Espinar." *American Ethnologist* 4:1:84–100.

Palmer, David Scott
1973 *Revolution from Above: Military Government and Popular Participation in Peru, 1968–72.* Latin Studies Program. Dissertation Series No. 47. Ithaca: Cornell University.

Palmer, David Scott, and Middlebrook, Kevin Jaj
1976 "Corporatist Participation and Military Rule in Peru." In *Peruvian Nationalism: A Corporatist Revolution.* edited by David Chaplin, pp. 428–53. New Brunswick, N.J.: Transaction Books.

Palomino Flores, Salvador
1971 "La Dualidad en la Organizacion Socio-Cultural de Algunos Pueblos del Area Andina." *Revista del Museo Nacional* 37:231–60.

Pasara, Luis
1970 *El Primer Año de Vigencia de la Ley de Reforma Agraria.* Lima: Centro de Estudios y Promoción del Desarrollo. Cuadernos Series.

Paulston, Rolland G.
1972 *Society, Schools, and Progress in Peru.* New York: Pergamon Press, Inc.

Payne, Arnold
1968 *The Peruvian Coup D'Etat of 1962: The Overthrow of Manuel Prado.* Washington: Institute for the Comparative Study of Political Systems.

Piker, Steven
1966 "The Image of Limited Good: Comments on an Exercise in Description and Interpretation." *American Anthropologist* 68:5:1202–11.

Pred, Allan
1967 *Behavior and Location: Foundations for a Geographic and Dynamic Location Theory.* Part 1. Lund Studies in Geography.

Series B. Human Geography No. 27. Sweden: The Royal University of Lund.

Quijano, Anibal
1965 "El movimiento Campesino del Perú y sus Lideres." *América Latina* 8:43–65.
1968 "Tendencies in Peruvian Development and in the Class Structure." In *Latin America: Reform or Revolution,* edited by James Petras and Maurice Zeitlin. Greenwich, Conn.: Fawcett Publications, Inc.

Ringlien, Wayne R.
1972 "Some Economic and Institutional Results of the Agrarian Reform in Peru." *Land Tenure Center Newsletter* 38:5–14.

Roberts, B., and Samaniego, C.
N.D. *The Significance of Agrarian Reform in the Highlands of Central Peru.* Manuscript.

Rogers, Everett M.
1969 "Motivations, Values, and Attitudes of Subsistence Farmers: Toward a Subculture of Peasantry." In *Subsistence Agriculture and Economic Development,* edited by Clifton Wharton, pp. 111–35. Chicago: Aldine Publishing Co.

Rogers, Everett M., with Svenning, Lynne
1969 *Modernization Among Peasants: The Impact of Communication.* New York: Holt, Rinehart and Winston.

Rubin de Celis, Emma
1977 *Las CAPS de Piura y sus Contradicciones.* Piura: Centro de Investigatión y Promoción del Campesinado.

Sabogal Wiesse, Jose R.
1969 "El subprefecto, los carneros, y Chacan." In *La Comunidad Andina Ediciones.* Especiales 51. Mexico D.F.: Instituto Indigenista Interamericano.

Schaedel, Richard P.
1972 *Variations in the Patterns of Contemporary and Recent Urban-Rural (Macro-Microsocietal) Linkages in Latin America.* Paper presented at the 40th Congress of Americanists, Rome.

Schultz, Theodore W.
1964 *Transforming Traditional Agriculture.* New Haven, Conn.: Yale University Press.

Skeldon, Ron
1974 *La SAIS como un modelo para Colonizacion: El Caso de la SAIS Túpac Amaru.* Lima: Centro Nacional de Capacitación e Investigación para la Reforma Agraria.
1976 "Population Migration and Regional Associations: An Interpretation." *Urban Anthropology* 5:233–52.
1977*a* "The Evolution of Migration Patterns during Urbanization in Peru." *The Geographical Review* 67:4:394–411.
1977*b* "Regional Associations: A Note on Opposed Interpretations." *Comparative Studies in Society and History* 19:4:506–10.

Sociedad Nacional Agraria
1963–1964 *Memoria.* Lima: Imprenta del Ministerio de Guerra.

Spaulding, Karen
1969 "Peru: the Military Managers." *Leviathan* (July-August) 1:4:-35–40.

Stacey, Margaret
1969 "The Myth of Community Studies." *British Journal of Sociology* 20:134–47.

Steward, Julian
1955 *Theory of Culture Change.* Urbana: University of Illinois Press.

Thorner, Daniel; Kerblay, Basile; and Smith, R. E. F., eds.
1966 *A. V. Chayanov: The Theory of Peasant Economy.* Homewood, Ill.: Richard D. Erwin.

Valderrama, Mariano
1976 *7 Años de Reforma Agraria Peruana: 1969–1976.* Lima: Pontificia Universidad Católica del Perú.

Van den Berghe, Pierre L., and Primov, George P.
1977 *Inequality in the Peruvian Andes. Class and Ethnicity in Cuzco.* Columbia, Mo.: University of Missouri Press.

Villanueva, Victor
1972 *El CAEM y la Revolucion de la Fuerza Armada.* Lima: Instituto de Estudios Peruanos.

Webster, Steven
1971 "An Indigenous Quechua community in Exploitation of Multiple Ecological Zones." *Actas y Memorias del XXXIX Congreso Internacional de Americanistas* 3:174–83. Lima.

Werlich, David P.
1977 "The Peruvian Revolution in Crisis." *Current History* 72:61–64 and 81–82.

Wolf, Eric
1956 "Aspects of Group Relations in a Complex Society. *American Anthropologist* 58:1065–78.
1966 *Peasants.* Englewood Cliffs, N.J.: Prentice-Hall, Inc.
1967 "Levels of Communal Relations." In *Handbook of Middle America,* edited by Manning Nash, Vol. 6. Austin: University of Texas Press.

Yepes del Castillo, Ernesto
1972 *Perú 1820–1920: Un Siglo de Desarrollo Capitalista.* Lima: Instituto de Estudios Peruanos.

Zaldivar, Ramon
1974 "Agrarian Reform and Military Reformism in Peru." In *Peasants, Landlords, and Governments: Agrarian Reform in the Third World,* edited by David Lehman, pp. 25–69. New York: Holmes & Meier Publishers, Inc.

Index